In Critical Condition

Representation and Mind
Hilary Putnam and Ned Block, editors

In Critical Condition

Polemical Essays on Cognitive Science and the Philosophy of Mind

Jerry Fodor

A Bradford Book
The MIT Press
Cambridge, Massachusetts
London, England

© 1998 Massachusetts Institute of Technology

All rights reserved. No part of this book may be reproduced in any form by any electronic or mechanical means (including photocopying, recording, or information storage and retrieval) without permission in writing from the publisher.

This book was set in Palatino on the Monotype "Prism Plus" PostScript Imagesetter by Asco Trade Typesetting Ltd., Hong Kong.

Printed and bound in the United States of America.

Library of Congress Cataloging-in-Publication Data

Fodor, Jerry A.
  In critical condition : polemical essays on cognitive science and
  the philosophy of mind / Jerry Fodor.
      p.   cm. — (Representation and mind)
  "A Bradford book."
  Includes bibliographical references (p.   ) and index.
  ISBN 0-262-06198-8 (hc : alk. paper)
  1. Philosophy of mind. 2. Cognitive science. I. Title.
II. Series.
BD418.3.F625   1998
128'.2—dc21
                                              98-21611
                                                CIP

A powerful idea communicates some of its power to the man who contradicts it. Partaking of the universal community of minds, it infiltrates, grafts itself onto, the mind of him whom it refutes, among other contiguous ideas, with the aid of which, counterattacking, he complements and corrects it; so that the final verdict is always to some extent the work of both parties to a discussion. It is to ideas which are not, strictly speaking, ideas at all, to ideas which, based on nothing, can find no foothold, no fraternal echo in the mind of the adversary, that the latter, grappling as it were with thin air, can find no word to say in answer.
Marcel Proust

Contents

Christopher Peacocke has been working out for some years and which is most explicitly set forth in his book *A Study of Concepts* (see chapter 3 of the present volume for an overview). Quite a lot of the philosophy I've done of late is an attempt to not be co-opted by Peacocke's project, while at the same time not getting lost. It turns out that's not so easy. I'm indebted to Peacocke for having provided a particularly bright light to steer by—and away from.

Chapter 1 was originally published in *The London Review of Books*, April 20, 1995. Chapter 2 was originally published in J. Tomberlin (ed.), *Philosophical Perspectives, 11, Mind, Causation, and World*, Ridgeview, CA. Chapter 3 was originally published in *The London Review of Books*, April 20, 1995. Chapter 4 was originally published in *Philosophical Issues*, Vol. 9 (E. Villanueva, ed.), Ridgeview Publishing Co., 1998. Chapter 7 was originally published in *The London Review of Books*, October 30, 1997. Chapter 8 was originally published in *The Times Literary Supplement*, August 25, 1995. Chapter 9 was originally published in *Cognition*, 1990, 35, 183–204. Chapter 10 was originally published in *Cognition*, 1996, 62, 109–119. (I'm grateful to Professor Brian McLaughlin for agreeing to my reprinting this paper.) Chapter 12 was originally published in *The London Review of Books*, Vol. 20, No. 12, 1997. Chapter 13 was originally published in *The London Review of Books*, Vol. 18, No. 23, 1996. Chapter 14 was originally published in *The London Review of Books*, Vol. 18, 1996. Chapter 15 was originally published in *Mind and Language*, 1996, Vol. 11, No. 3, pp. 246–262. Chapter 17 was originally published in *The London Review of Books*, Jan. 22, 1998. I am grateful to all of these sources for their kind permission to publish this material here.

A word about style: many of these papers were originally intended for an audience that was either lay or interdisciplinary. In those, hard core philosophy is kept to a minimum. There are also a lot of jokes, for which I am sometimes taken to task. I do admit to finding many of the views that I argue against pretty funny. I find many of my own views pretty funny too. Nietzsche is righter than Brecht: Sometimes the man who laughs *has* heard the terrible news.

Preface

Philosophy, like piloting, is mostly about figuring out where you are. The basic principle of both is much the same: Find an object whose position is known and locate yourself with respect to it. Typically, in both cases, the object in question is somewhere that you do not wish to be: On a rock, on a shoal, at the edge of the channel, or half a mile inland from the shore. So the trick is to get close enough to recognize the thing and to figure out what it means, but not so close that it swallows you up. Thus Aristotle with respect to Plato, Kant with respect to Hume, Descartes with respect to Newton, and me with respect to many a bell buoy on many a summer Sunday afternoon.

The difference is that whereas navigators don't often argue with their landmarks, philosophers hardly ever do anything else. Hence this collection. Each of these essays reacts to a position in philosophy or cognitive science (in the present context the distinction is hardly worth drawing) that I'm pretty sure I don't want to occupy, but that I'm also pretty sure needs to be marked on the charts and, as sailors say, "honored." I think that not getting wrecked requires avoiding all of them at once. And I think that there's a way to do that: What's required is a mix of intentional realism, computational reductionism, nativism, and semantic atomism, together with a representational theory of mind. I've talked and written about all that elsewhere, however; in these polemical essays the positive program is present largely by implication. Very roughly, and just for orientation: The views examined in Part I bear on questions of realism and reductionism, and those in Part II mostly concern semantic atomism. Part III is about what, if any, sense can be made of the idea that mental processes are computational. Part IV is about what, if any, constraints evolutionary theory imposes on semantics and the philosophy of mind.

Many of these papers are book reviews and the targets are identified by the titles. Others are responses to journal articles which are likewise eponymous. The apparent exceptions are chapters 4 and 5, the ones about recognitional concepts. In fact, however, these were prompted by, and constitute an extended commentary on, a philosophical project that

Part I
Metaphysics

Chapter 1

Review of John McDowell's *Mind and World*

Mind and World collects McDowell's 1991 John Locke lectures, with some of his afterthoughts appended. Until you're halfway through, the book seems to be about a relatively technical question: What's the relation between the mind's contribution to perceptual experience and the world's? A deep worry, to be sure; but one that's maybe best left for epistemologists to thrash out. In chapter 4, however, McDowell achieves a coup de theatre. The angle of view widens unexpectedly, and the specifically epistemological issue is seen to be a microcosm of some of modern philosophy's most characteristic concerns: How should we construe the relation between the realm of reason and the realm of natural law? How can it be that we are both physical objects and thinking things? How can conceptualization be free and spontaneous if the mind is a mechanism? Why, in short, isn't "rational animal" an oxymoron? This is a fine expository contrivance, and the framing of the issues that it permits is deeply illuminating. McDowell's book finds new ways to register a number of our deepest perplexities.

Which, however, is not to say that I believe a word of it.

A dialectic between two different and opposed conceptions of naturalism—in particular, of a naturalistic account of rationality—is working itself out in *Mind and World*. There's the reductionist version (McDowell calls it "bald" naturalism; "scientism" is another pejorative that is currently in fashion). And there's the kind of naturalistic pluralism that McDowell himself is striving for. Very roughly, the distinction is between the tradition that runs from Kant through the positivists to the likes of Dewey and Quine, and the tradition that runs from Kant through the Hegelians to Wittgenstein, Ryle, and Davidson (and Hilary Putnam since he left MIT for Harvard).

It's hard to be articulate about this disagreement; we're very close to the edge of what we know how to talk about at all sensibly. But, for reductionists, the world picture that the natural sciences lay out has a sort of priority—sometimes viewed as metaphysical, sometimes as methodological, sometimes as ideological, sometimes as all of these at once—to which other discourse is required to defer insofar as it purports to speak literal truths. Conflicts between the scientific image and, for example, the claims that moral theories make, or theories of agency, or theories of mind, are real possibilities, and if they arise, science is privileged—not because the "scientific method" is infallible, but because the natural realm is the only realm there is or can be; everything that ever happens, including our being rational, is the conformity of nature to law. And our science is the best attested story about the conformity of nature to law that we know how to tell. Accordingly, the philosophical problems about mind and world have to be situated within the general scientific enterprise, if literal truth is what philosophers aim for. What our rationality consists in is an open question, apt for a kind of inquiry that is empirical and metaphysical at the same time; as, indeed, scientific inquiry is wont to be.

For pluralists, however, the situation presents itself quite differently. There are lots of more or less autonomous varieties of discourse (or world views, or language games, or forms of life, or paradigms), and the critique they are subject to is largely from inside and in their own terms. As McDowell puts it, "Even a thought that transforms a tradition must be rooted in the tradition that it transforms" (187; all references in this paper are to McDowell, 1994). Accordingly, the natural scientist's activity of limning a normless and otherwise "disenchanted" natural order is just one way of world-making among others. For the epistemologist's purposes, in contrast to the scientist's, the normative character of rational assessment is given a priori; to that extent, we already know what the essence of rationality is. The problem is to find a place for it outside what the natural sciences take to be the natural order, but to do so without, as McDowell sometimes says, thereby making rationality look spooky.

Pretty clearly McDowell thinks that bald, reductive naturalism isn't seriously an option, so it's going to be pluralism or nothing if the integrity of the rational is to be sustained. Just why he thinks this is less clear. Circa page 74 there's some moderately loose talk, with a nod to Donald Davidson, about the "constitutive principles" of rationality being such that "the logical space that is the home of the idea of spontaneity cannot be aligned with the logical space that is the home of ideas of what is natural in the relevant sense [viz. with] the characteristically modern conception according to which something's way of being natural is its position in the realm of law." The unwary reader might suppose from this that somebody has actually argued conclusively that the reductionist program can't be carried through, hence that either the mutual autonomy of the natural order and the rational order is somehow guaranteed, or else there is no such thing as rationality; the latter outcome being, to use a favorite epithet of McDowell's, "intolerable."

In fact, of course, nothing of the sort has been shown, nor will be. Bald naturalism may, for all philosophers know, be viable after all. That may strike you as comforting if, like me, you find McDowell's efforts to formulate a pluralistic alternative rather less than convincing.

I guess I am a hairy naturalist: Though I do agree that the problems about mind and world are a lot harder than reductionists have sometimes supposed, I also think that an adequate and complete empirical psychology would, ipso facto, tell the whole, literal truth about the essence of the mental. Science discovers essences, as Kripke once remarked. So, if it's literally true that rationality, intentionality, normativity, and the like belong to the mind essentially, then they must all be phenomena within the natural realm that scientists explore. McDowell comments, sort of in passing, that "cognitive psychology is an intellectually respectable discipline ... so long as it stays within its proper bounds." That, however, is truistic, proper bounds being what they are. The serious question is whether there is anything about mentality that can be excluded from scientific inquiry a priori; anything, anyhow, that claims to be both of the essence and literally true.

Consider, for example, the epistemological question that McDowell starts with. You might have thought that seeing a tree goes something like this: Light bounces off the tree and affects your eyes in ways that determine the sensory content of your experience. Your mind reacts by inferring from the sensory content of your experience something about what in the world must have caused it. The upshot, if all goes well, is that you see the world as locally entreed.

So, then, perception is a hybrid of what the senses are given and what the mind infers. The process is causal through and through: It's part of psychophysics that encounters with trees bring about the kinds of visual sensations that they do. And it's part of cognitive psychology that the visual sensations that encounters with trees provoke occasion the perceptual inferences that they do in minds with the right kinds of history and structure. To be sure, the story as I've just told it is bald and insufficiently detailed; but ironing out its wrinkles is what perceptual psychologists are paid to do, and my impression is that they're getting along with the job pretty well. So what, exactly, is supposed to be wrong?

McDowell's answer is entirely characteristic: "The trouble about the Myth of the Given is that it offers us at best exculpations where we wanted justifications.... [T]he best [it] can yield is that we cannot be blamed for believing whatever [our experiences] lead us to believe, not that we are justified in believing it" (13). That is: McDowell has in mind a certain account of what the rationality of a perceptual judgment consists in, and that account isn't satisfied if the world's contribution is merely to provide the judgment with the right sort of etiology. The empirical story says that what the world gives beliefs come from the sensory states that it causes us to have. But that won't do since sensations aren't reasons; "nothing can count as a reason for holding a belief except something else that is also in the space of concepts ..." (140). Period. To have to tinker with the epistemological story—to have to reconsider what rational judgment is in light of a merely empirical theory about the conditions under which it can be achieved—would strike McDowell as, well, intolerable.

So McDowell is committed to a view that might well strike you as hopeless on the face of it; he needs to be a naturalist and a dualist at the same time. On one side, what the world contributes to perception must be something that one can think. It must have, so to speak, the kind of structure that thoughts have, for only then does the mental process that gets from experience to judgment count as rational by McDowell's criteria. But, on the other side, McDowell is quite aware that the world isn't any kind of text, and that idealism has to be avoided. "In a common medieval outlook, what we now see as the subject matter of natural science was conceived ... as if all of nature were a book of lessons for us.... It is a mark of intellectual progress that educated people cannot now take that idea seriously" (71). One is therefore not to argue as idealists might wish to do: that if the world's contribution to perception has to be something that's thinkable, then since all that is thinkable is thoughts, the world must be made of thoughts if we are to be able to perceive it.

These views are, however, alternatives that a lot of philosophers hold to be exhaustive. Granting something unconceptualized that is simply Given to the mind in experience has generally been supposed to be the epistemological price one has to pay for an ontology that takes the world to be not itself mind-dependent. The epistemological passages in McDowell's book struggle to find space between these options. I'm not at all sure they do. I'm not at all sure that there is any space to find. From the comfortable perspective of my kind of naturalist, anyhow, the whole undertaking seems to carry contortion beyond necessity. Laocoön looks a little comical if there's no snake. Until the second appendix, McDowell doesn't even consider taking what is surely the easy way out: Maybe sometimes exculpation *is* justification and is all the justification that there is to be had.

Why not, after all? If the situation is that I can't but believe that I'm looking at a tree, and if, in that situation, it's the case that I am looking at a tree, and if there is a workable account of why, in such situations, I reliably come to believe that there's a tree that I'm looking at (viz., because they are situations where trees cause the kind of sensations that cause minds like mine to think that they are seeing trees), why isn't that good enough for my judgment that I'm seeing a tree to make it into the Realm of Reason? Why, in short, mightn't fleshing out the standard psychological account of perception itself count as learning what perceptual justification amounts to?

When McDowell does finally consider this kind of option, he switches ground disconcertingly. Suddenly, the worry isn't that causation provides for exculpation but not for justification; rather, it's that the justification it provides for needn't constitute *"a subject's reasons for* believing something" (*sic*, 161). This all goes very fast, and I doubt that it actually amounts to much. When, in the situation imagined, I come to believe that I see a tree, *my* reason for believing there's a tree is that I see it; and my reason for believing that I see it is that I do. Why do I think I see a tree there? Well, why on earth should I not? Just look at the thumping great thing! Mcdowell's sense of how justification actually goes occasionally strikes me as excessively earnest.

This easy path having once been eschewed, the hard path proves to be very hard indeed: "If we can rethink our conception of nature so as to make room for spontaneity, even though we deny that spontaneity is captured by the resources of bald naturalism, we shall by the same token be rethinking our conception of what it takes for a position to be called 'naturalism'" (77). More bluntly: The cost of McDowell's a priorism is that he has to be some sort of dualist; not necessarily the Cartesian sort, who thinks that there are immaterial *things*. But, quite likely, the kind of faculty dualist who is, willy-nilly, landed with occult powers. Having situated the rational (and the ethical, and a lot else that we care about) outside the realm of law, McDowell needs to face the embarrassing question how, by any natural process, do we ever manage to get at it?

He needs to, but in fact he doesn't. "When we are not misled by experience, we are directly confronted by a worldly state of affairs itself, not waited on by an intermediary that happens to tell the truth" (143). But if you want to hold that states of affairs themselves are what veridical perception works on, you need a story about how unmediated cognitive connections to states-of-affairs themselves might be achieved. Likewise, and more so, in the nonperceptual cases, where the objects of cognition are normative or otherwise intentional aspects of things: how do we get at those if they aren't in the natural order? Maybe better, how do they get at us? *How can what is not in the realm of law make anything happen?*

Here's what I take to be McDowell's answer: "We need to bring responsiveness to meaning back into the operations of our natural sentient capacities as such" (77); as he sometimes puts it, we need somehow to think of the mind as "resonating" to rational relations. Or consider: "the rational demands of ethics are not alien to the contingencies of our life as human beings.... [O]rdinary upbringing can shape the actions and thoughts of human beings in a way that brings these demands into view" (83). But "bringing into view" is a metaphor; only what is in nature can literally be viewed. And "resonating" is also a metaphor; only what is in nature can be literally attuned to. The trouble with "putting responsiveness to meaning back into the operations of our natural sentient capacities as such" is that nobody has the foggiest idea of how to do so unless both are contained in the natural order.

The forms of human sentience resonate, as far as anybody knows, only to aspects of the "disenchanted" world. Mere exhortation won't fix that. "We tend to be forgetful of ... second nature. I am suggesting that if we can recapture that idea, we keep nature as it were partially enchanted, but without lapsing into prescientific superstition or a rampant platonism" (85). "[O]nce we allow that natural powers can include powers of second nature, the threat of incoherence disappears" (88). Second nature is what we get when "our *Bildung* actualizes some of the potentialities we are born with; we do not have to suppose it introduces a nonanimal ingredient into our constitution." But the question arises how second nature, so conceived, could itself be natural. It's not good enough for McDowell just to say that it is and you can get some at the *Bildung* store; he has to say how it could be, short of spooks. Otherwise, why is McDowell's kind of dualism to be preferred to Descartes's?

What McDowell has to pay for demanding that his account of perception conform to the a priori constraints that his normative epistemology imposes is that he leaves us with no idea at all how perceiving could be a process in the world. Since I, for one, can't imagine how a faculty could resonate to meanings "as such," this seems to me not to be worth the cost. I'm afraid the bottom line is there is no room where McDowell wants to wiggle; a dualistic naturalism isn't in the cards. If that's right, then epistemology needs to bend and McDowell will have to cool it a little about justification. Justification can't require what can't happen, on pain of there not being any; and whatever happens, happens in the realm of law.

Ever since Descartes, a lot of the very best philosophers have thought of science as an invading army from whose depredations safe havens have somehow to be constructed. Philosophy patrols the borders, keeping the sciences "intellectually respectable" by keeping them "within ... proper bounds." But you have to look outside these bounds if what you care about is the life of the spirit or the life of the mind. McDowell's is as good a contemporary representative of this kind of philosophical sensibility as you could hope to find. But it's all wrong headed. Science isn't an enemy, it's just us. And our problem isn't to make a place in the world for the mind. The mind is already in the world; our problem is to understand it.

Chapter 2

Special Sciences: Still Autonomous after All These Years (A Reply to Jaegwon Kim's "Multiple Realization and the Metaphysics of Reduction")

The conventional wisdom in philosophy of mind is that "the conventional wisdom in philosophy of mind [is] that psychological states are 'multiply realized' [and that this] fact refutes psychophysical reductionism once and for all" (Kim, 1993, 309. All Kim references are to this paper except as noted). Despite the consensus, however, I am strongly inclined to think that psychological states are multiply realized and that this fact refutes psychophysical reductionism once and for all. As e. e. cummings says somewhere: "*nobody* loses *all* of the time."

Simply to have a convenient way of talking, I will say that a law or theory that figures in bona fide empirical explanations but that is not reducible to a law or theory of physics is ipso facto *autonomous*, and that the states whose behavior such laws or theories specify are *functional* states. (In fact, I don't know whether autonomous states are ipso facto functional. For present purposes all that matters is whether functional states are ipso facto autonomous.) So, then, the conventional wisdom in the philosophy of mind is that psychological states are functional and the laws and theories that figure in psychological explanations are autonomous.[1] (Likewise, and for much the same reasons, for the laws, theories, etc. in other "special" [viz. nonbasic] sciences.) The present discussion undertakes to defend this consensus view against a philosophical qualm that is the main moral of Jaegwon Kim's paper "Multiple Realization and the Metaphysics of Reduction."

Kim says that he's prepared to agree (at least for the sake of the argument) that:

1. Psychological states are typically multiply realized (MR); and
2. MR states are ipso facto unsuitable for reduction.

But Kim thinks philosophers haven't gotten it right about *why* MR states are ipso facto unsuitable for reduction. Once they do, Kim says, they'll see that the moral of 1 and 2 isn't, after all, that psychology is autonomous. Rather, it's that quotidian psychological states aren't *reducible* because they aren't *projectible*. Unprojectible states are, by definition, not the subjects of a possible science; they aren't bona fide kinds and they can't appear in bona fide nomological explanations. A fortiori, terms that express psychological states are not available for incorporation in "bridge laws" or in (metaphysically necessary) property identities. This is all, of course, contrary to what a lot of philosophers, to say nothing of a lot of psychologists, have hitherto supposed.

Caveat: It turns out, according to Kim, that some very finely individuated psychological states are "locally" reducible after all; but that's because these states are, by assumption, not MR. More on this later; I mention it here just to make clear that Kim is *not* claiming that psychological states are unprojectible *qua psychological* (or qua intentional), but only that they are unprojectible qua not local. Contrast (e.g.) Davidson (1980).

Now, I think Kim is quite right that what moral you should draw from MR states not being reducible depends on whether whatever it is that makes them not reducible also makes them not projectible. But I think that the diagnosis of the irreducibility of MR states that Kim offers is wrong, and that the right diagnosis supports the standard view: "pain," "believes that P," and the like express real states, about which all the available evidence suggests that there are real laws.[2] If I'm right that such states are projectible but not reducible, then it follows that psychological laws are autonomous after all.

So much for strategy; now for tactics. Kim's polemical method is to pick an apparently untendentious (and, in particular, a *non*psychological) MR state and explain why it is unprojectible qua MR. He will then argue that since beliefs, pains, and the like are MR by assumption (1), they must be unprojectible for the same reasons that his MR paradigm is. Tit for tat: I'll argue both that Kim's diagnosis of the unprojectibility of his MR paradigm is wrong and that the supposed analogy between his MR paradigm and pains, beliefs, and the like is spurious.

More caveat: In order not to be always writing "pains, beliefs, and the like" or "quotidian psychological states," I'll follow Kim's practice and take pain as my working example of a quotidian psychological state. In particular, I'll assume that, if there are any psychological laws at all, then probably there are psychological laws about pain. In fact, however, pain isn't really a happy choice for a working example, since though I take it to be quite a plausible candidate for projectibility, pain is notoriously a very bad candidate for being MR, hence for functional analysis. That, however, is stuff for a different paper. (See, e.g., Block and Fodor, 1972; Block, 1978.) My present brief is just that what Kim says is wrong with functionalism actually isn't; so the autnomy thesis is, to that extent at least, not in jeopardy. For these purposes, pain will do.

So, finally, to work.

Jade: We are told, Kim tells us, that jade "is not a mineral kind, contrary to what was once believed; rather, jade is comprised of two distinct minerals with dissimilar molecular structure, *jadeite* and *nephrite*" (319). Geological unsophisticate that I am, I shall often call these two minerals *jadeA* and *jadeB* in the discussion that follows. It won't really matter which is which, so you needn't bother to keep track.

Kim thinks that, because of these facts about jadeite and nephrite, jade is paradigmatically MR. Kim also thinks that since jade is paradigmatically MR, "jade" is ipso facto unprojectible. And, a fortiori, (3) isn't a law.

> 3. Jade is green

I don't actually care much whether (3) is a law or even whether "jade" is projectible. But I am going to deny that jade is paradigmatically MR, and I am also going to deny that "jade" is unprojectible for the reason that Kim says it is.

So then, what, according to Kim, is wrong with (3)? Well, "[l]awlike generalizations ... are thought to have the ... property [that] observation of positive instances, Fs that are Gs, can strengthen our credence in the next F's being G." In short, real laws are confirmed by their instances, but (3), according to Kim, is not: "[W]e can imagine this: on reexamining the records of past observations, we find, to our dismay, that all the positive instances of (3) ... turn out to have been samples of jadeite, and none of nephrite! If this should happen ... we would not ... continue to think of (3) as well confirmed.... But all the millions of green jadeite samples *are* positive instances of (3).... The reason [that (3) is not confirmed by them] is that jade is a true disjunctive kind, a disjunction of two heterogeneous nomic kinds which, however, is not itself a nomic kind" (320)

Now notice, to begin with, that the thought experiment Kim proposes doesn't really make the case that he wants it to. To be sure, if we discovered that all our samples of green jade are samples of green jadeA, that would lead us to doubt that (3) is well confirmed. Not, however, because it would be a reason to think that (3) doesn't "pass the projectibility test" (320); but rather, because it would show that we've made a sampling error in collecting the data. The point is: *Anybody* can make a sampling error, *whether or not* the hypothesis he's trying to confirm is projectible.

Suppose we've been considering whether oak trees shed their leaves in winter; and suppose it turns out, on reexamining the records, that our positive instances are all observations of oak trees on the north side of hills. Then we would no longer think of the generalization about oak trees losing their leaves in the winter as unambiguously well confirmed; oak data confirm oak generalizations only if they are an *unbiased* sample of the oak population, which, on the current assumption, our data aren't. Maybe, in the present case, the generalization that the instances really confirm is that *oak trees on the north side of hills* lose their leaves in winter. But notice that discovering a sampling error of this sort would be no reason at all for doubting that *oak tree* is a kind. Rather, the worry would be that maybe *oak tree on the north side of a hill* is a kind too. If it is, then our data are equivocal between two perfectly okay, projectible hypotheses: the one that goes *blah, blah … oak trees …*, and the one that goes *blah, blah…. oak trees on the north sides of hills….* When we discover the sampling error, we regard neither generalization as unequivocally confirmed by data that are instance of both, and this is precisely because the data *are* instances of both. *The sampling error means that the data are equivocal, not that the hypotheses are unprojectible.* There is, to be sure, something wrong with (3); something that makes it not a law. But what's wrong with (3) isn't that biased samples fail to confirm it. *Biased* samples don't confirm *anything*.

So Kim's thought experiment shows nothing that is to the point. However, he doesn't really need the thought experiment to run his argument. Instead, he could run his argument like this:

Although green JadeA samples tell us that jadeA is green, and green jadeB samples tell us that jadeB is green, green jadeA samples tell us nothing at all about the color of jadeB and green jadeB samples tell us nothing at all about the color of jadeA. That's because, though it's true that jadeA is green iff jadeB is green, that it is true is merely accidental given the presumed facts about the structural heterogeneity of nephrite and jadeite. Analogously: Since jadeite is green and grass is green, jadeite is green iff grass is. But it doesn't follow that evidence about the color of grass bears at all on hypotheses about the color of jadeite or vice versa.

But, now, according to the functionalist orthodoxy, pain is just like jade, isn't it? So, then:

> Why isn't pain's relationship to its realization bases, N_h, N_r, N_m analogous to jade's relation to jadeite and nephrite? If jade turns out to be nonnomic [i.e., not projectible] on account of its dual "realizations" in distinct microstructures, why doesn't the same fate befall pain? After all, the group of actual and nomologically possible realizations of pain, as they are described by the MR enthusiasts with such imagination, is far more motley than the two chemical kinds comprising jade…. We put the following question to Fodor and like-minded philosophers: If pain is nomically equivalent to N, the property claimed to be wildly disjunctive and obviously nonnomic, *why isn't pain itself equally heterogeneous and nonnomic as a kind?* (323; emphasis Kim's)

I expect this question is meant to be rhetorical, but I think that I'll answer it all the same.

Rebuttal: There is, I think, a sort of polemical standoff at this point in the discussion. Kim is quite right that jade is "a true disjunctive kind, a disjunction of two heterogeneous nomic kinds which, however, is not itself a nomic kind" (12). But, from the functionalist's perspective, to say that jade is a disjunctive kind because jade generalizations aren't confirmed by their instances is to put the epistemological cart before the ontological horse. The right story, according to the functionalist, goes the other way around. What makes jade a disjunctive kind (in the, as we're about to see, special sense that is germane to whether it's projectible) is that *there are no general empirical truths about jade as such*; a fortiori, there are no such truths for samples of jade, as such, to confirm. Whatever is reliably true of jade is so either because it is reliably true of *jadeA* as such or because it is reliably true of *jadeB* as such.

No doubt there are those of you who are suspicious of "as such's" as such; but, in fact, none is required to make the present point. What's needed is just the distinction between a multiply based property that is *disjunctive* and a multiply based property that is disjunctively *realized*. To wit: A multiply based property is disjunctive iff it has no realizer in *any* metaphysically possible world that it lacks in the *actual* world. Jade is disjunctive in that the only metaphysically possible worlds for jade are the ones which contain either jadeA or jadeB or both. By contrast, multiply based properties that are disjunctively *realized* have different bases in different worlds. Pain is disjunctively realized in that there's a metaphysically possible, nonactual, world in which there are silicon-based pains.

That is to say, in effect, that being jade *just is* being jadeA or jadeB. Whereas it would be simply question-begging of Kim to hold that being pain is the property of being one or another of pain's realizers. Functionalists claim that pains and the like are higher-order, relational properties that things have in virtue of the pattern of causal interactions that they (can or do) enter into. *If so, then pains, though multiply based, are not disjunctive but MR*. To repeat: Though Kim *says* that he concedes that psychological properties are MR, that's only because he isn't distinguishing *being* MR (like pain) from *being disjunctive* (like jade). But it's exactly the distinction between *disjunctiveness* and disjunctive *realization* that functionalists are insisting on when they say that pain states are nomologically homogeneous under their functional description *despite the physical heterogeniety of their realizers*. You can't (and Kim can't) refute this claim just by defining "disjunctive kind" so that it isn't true.

This is, as I say, a sort of polemical standoff. The functionalist assumes that there are laws about pains "as such"; so he infers that, though pain is multiply based, it is *not* (merely) disjunctive. So he infers that pain is unlike jade in the respects that are relevant to the question of projectibility. Kim, going the other way around, assumes that pain is (merely) disjunctive, hence that it is relevantly similar to jade, hence that there aren't any laws about pain. The real issue—the one that Kim begs with his appeal to the jade example—is whether there *is* a difference between a multiply based property's being disjunctive and its being MR; and, if so, what the difference is and whether it matters to projectibility. Kim almost sees this in the closing sections his paper. But then he gets it wrong—fatally, in my view. "Is jade a *kind*? We know it is not a mineral kind; but is it any kind of a kind.... There are certain shared criteria, largely based on observable macroproperties of mineral samples (e.g. hardness, color, etc.), that determine whether something is a sample of jade.... What all samples of jade have in common is just these observable macrophysical properties that define the applicability of the predicate 'is jade'" (24). This, I say, is just wrong; and resolving the metaphysical issues about projectibility that Kim has raised turns on seeing that it is.

Suppose that, puttering around in the cellar one day, you succeed in cooking up a substance—out of melted bottle glass, let's say—that is, for all macroscopic purposes, indistinguishable from jade: For example, it's as similar in color to jadeA and jadeB as they typically are to one another; its hardness falls at about the right place between talc and diamond on the scratch test; it cracks along the right sort of cleavage planes; it weighs about the same as jade per unit volume, and so forth. Have you, then, *created jade*? Oh frabjous day! Oh joy that alchemists never knew! Oh (in particular) riches running wild!

Not on your Nelly. What you've got there isn't jade; it's just melted bottle glass. Melted bottle glass maybe counts as *artificial* jade in the sort of case that we've imagined; but *do not try to sell it as the real stuff.* You will find, if you do, that fanciers of jade are not amused. They will call you unkind things like *thief* and *fraud*; and if they catch you, they will lock you up. Pace Kim, being jade is *not* relevantly like having a functional (i.e., MR) property; if it were, you could make new jade by mimicking the macroscopic properties that jadeA and jadeB share (the ones which, according to Kim, "determine whether something is a sample of jade, or whether the predicate 'is jade' is correctly applicable to it"). But you can't make jade that way. If you want to make some jade, you have to make either some jadeite or some nephrite; just as, if you want to make some water, you have to make some H_2O. That's because *jade is jadeite or nephrite is metaphysically necessary*, just like *water is H_2O.*[3] As with most of the metaphysical claims one comes across these days, the one that I just made relies for its warrant on a blatant appeal to modal intuitions. But I think the modal intuitions that I'm mongering are pretty clearly the right ones to have. If you don't share them, perhaps you need to have yours looked at.

Now compare pain according to the functionalist persuasion. Suppose you should happen, one day down in the cellar, to throw together a robot, among whose types of internal states there's one that is, under functional description, about as similar to my pains as my pains are to yours. Then, functionalists say, the thing that you've created *has pains.* Not *artificial* pains, not pain *simulations,* not *virtual* pains, but *the real things.* Even if, as we may suppose, what you've cooked up is a silicon robot that's made out of melted bottle glass. My point is not, I emphasize, to claim that the functionalist is right to be of this persuasion. For reasons I've elaborated elsewhere, I doubt that pain is a functional kind, or that beliefs-that-P are either. What's at issue, remember, isn't whether functionalism is true of the mental; it's only whether there is a plausible functionalist response to the challenge that Kim has laid down: "Why isn't pain's relationship to its realization bases ... analogous to jade's relationship to jadeite and nephrite?" Reply: There's a difference between being a functional property (being multiply realized) and being a disjunctive property. Being jade, according to the geologists, is an example of the latter; being pain, according to the psychologists, is an example of the former. So there is, thus far, nothing to warrant an inference from the unprojectibility of "jade" to the unprojectibility of "pain." Kim is quite right to emphasize that "the group of actual and nomologically possible [and epistemically and metaphysically possible] realizations of pain, as they are described by the MR enthusiasts with such imagination is far more motley than the two chemical kinds compromising jade." What he's missed is that *for that very reason* the pain/jade analogy is flawed and won't do the polemical work that he wants it to.

If there is after all plausibly a difference between the relation between pain and its realizers, on the one hand, and the relation between jade and its realizers, on the other, then it's not patently irrational of a functionalist to agree that "jade" isn't projectible while continuing to claim that "pain" is. But what is not patently irrational may nonetheless be unreasonable and obstinate. Granted that pain and jade differ in the ways that we have been exploring, it remains open to Kim to wonder why that kind of difference should matter so much. How is it that MR properties are projectible but disjunctive properties aren't? Or, to put the same question slightly differently: Functionalists are required to deny that pain is *identical to* the disjunction of its realizers. The reason functionalists are required to deny this is that it's part of their story that the property realized, *but not the disjunction of its realizers,* is projectible. And the reason functionalists have to say *that* is that *otherwise multiple realization wouldn't be an argument against reduction:* What is supposed to make the case for the autonomy (unreducibility) of functional laws is that there are no laws about the disjunction that realizes a functional state *even if there are laws about the functional state itself*. So then functionalists must themselves think that the disjunction of realizers of even bona fide, projectible MR states aren't themselves projectible. What justifies a functionalist in making this claim?

Or, to put the question yet another way: Functionalists hold that the biconditionals that connect functional properties with their realizers aren't *laws* (a fortiori, that they aren't "bridge" laws). They *can't* be laws because the realizers of functional states are, by assumption, disjunctive; and disjunctive properties are supposed to be ipso facto not projectible. Thus, in one of his earliest functionalist papers, Putnam remarks that rendering realizers projectible by "defining the disjunction of two [realizing] states to be a single 'physical-chemical state' ... is not a metaphysical option that can be taken seriously" (Rosenthal, 1991, 201). I think Putnam is right that it's not. But the question *why* it isn't remains.

Here's where we've got so far: Functionalists agree with Kim that there are no laws about disjunctive properties, whether the disjunctions are metaphysically open like *being in pain*, or metaphysically closed like *being jade*. But, according to functionalists, there can be a projectible property that is coextensive with an open disjunction (e.g., being in pain); whereas, in the case of a closed disjunction, there can't be. So, what, according to the functionalist, accounts for this asymmetry? In one form or another, that question keeps coming up. It needs an answer.

Here's a possible story: Suppose, for a moment, that metaphysically open disjunctive properties are projectible after all. Still, there's a difference between the case where a disjunction appears *only* in bridge biconditionals, and the case where it is, as it were, independently certified because it also occurs in laws at its own level. It might be quite reasonable to hold that disjunctions are bona fide in bridge laws *only* when their projectiblity is independently certified in this way. That would distinguish "real" bridge laws from those containing formulas that are not only disjunctive, but also *gerrymandered*; the latter being cases where, intuitively, *all* that the disjuncts have in common is that they realize some higher level state.

That was, in effect, the view I took in an earlier paper (also called "Special Sciences," Fodor, 1974). The objection I voiced there was not to the projection of disjunctions—open or closed—as such, but rather to a theory of interlevel nomic relations that fails to distinguish real reductions from gerrymanderings. "Type" physicalism is supposed, by general consensus, to be stronger than "token" physicalism; stronger, that is, than the mere claim that all mental states are necessarily physically instantiated, however unhomogeneously. I suggested that the difference is that type physicalism, but not token physicalism, requires bridge laws that really are laws, namely, ones that contain predicates that are independently certified as projectible because they are independently required to state *intralevel* laws. By this "no gerrymandering" criterion, it is empirically plausible that the multiple realizers of functional states are often not projectible, either in psychology or elsewhere.

I still think that's perfectly all right, as far as it goes. In fact, I think it's quite attractive since it fends off the following dilemma. I suppose a functionalist might wish to admit that there are nomologically necessary constraints on what sorts of things can be realizers of pains (or can openers or whatever). That is, for each functional property P, there must be *some* disjunction of realizers (call it "#") such that "*all Ps are $R1$, or $R2$, or $R3$ ... #*" is nomologically necessary[4]; and even if we don't (or can't) know what # is, presumably God can (and does). But if "all Ps are $R1$, or $R2$, or $R3$ ... #" is nomologicaly necessary, it look likes P is reducible after all since it looks like "all Ps are $R1$, or $R2$, or $R3$ ... #" is a (bridge) law; presumably, a law is just a universal conditional that's true in all nomologically possible worlds. But the argument for the autonomy of metaphysically open MR states depends on their *not* being lawfully related to their realizers. So it looks like that argument must be unsound.

The way out of this for functionalists is to require that bridge laws be not just nomologically necesary but also that they be *not gerrymandered*, as per above. What's wrong with "all Ps are $R1$, or $R2$ ..." isn't that it lacks a nomologically necessary closure; "#" is one by assumption. What's wrong is rather that *the nomological closure is gerrymandered*. That is, the predicate "$R1$, or $R2$, or ... #" isn't "independently certified." That is, it doesn't occur in any *proper* ("single level") laws. Since "$R1$, or $R2$, or ... #" isn't independently certified, "all Ps are $R1$, or $R2$, or ... #" isn't a bridge law and (ceteris paribus) P isn't reducible after all. On this account, the constraints on bridge laws are *stronger than* (in effect, they include) the constraints on proper (single-level) laws. This difference is what underlies the intuition that type materialism comes to more than just the claim that it's nomologically necessary that every nonbasic property be physically realized.

As I say, I still like that story; but I admit that there may be more to be said. It's clearly Kim's intuition that there's something wrong with multiply based kinds as candidates for projectibility *whether or not* they are gerrymandered. That is, that you aren't allowed multiply based kinds either in inter- or in intralevel laws. For reasons we've already discussed, I don't accept Kim's account of why this prohibition holds; on my view, his diagnosis depends on his failure to distinguish closed disjunctions from open multiple realizations. But I'm prepared to take it seriously that maybe disjunctions are, as such, bad candidates for projection. And as we've just seen, even functionalists have to claim, at a minimum, that

nondisjunctive functional states are ipso facto better candidates for projection than are their open disjunctive realizers; otherwise open disjunctive laws about the latter would compete with nondisjunctive laws about
the former. So, then, *just what is wrong with projecting disjunctions?*

It's not hard to see why it's so plausible that there can't be laws about
closed disjunctions. By assumption, if P is the closed disjunction $F \vee G$,
then it is metaphysically necessary that the properties a thing has qua P
are either properties it has qua F or properties it has qua G; and, of course,
this includes projectible properties inter alia. That's why, if being jade
really is a closed disjunctive property (if being jade is just being jadeite or
nephrite) then *of course* there are no laws about being jade "as such"; all
the jade laws are ipso facto either jadeite laws or nephrite laws.

But it's not equally obvious why (or whether) there can't be laws
about *open* disjunctions as such. In fact, I think there are depths here to be
plumbed.

One might, for starters, try denying that open *formulas*—ones that
contain ellipses essentially—succeed in expressing properties at all; in
which case, their failure to express *projectible* properties would not be
surprising. But I propose to assume without argument that a predicate like
"is $R1 \vee R2 \ldots$" does pick out a corresponding property, namely, the
property of being $R1 \vee R2\ldots$. It is, remember, common ground between
Kim and the functionalists that mental states are multiply realized. So both
are committed to some sentence of the form "pain is $R1 \vee R2 \ldots$" being
true, and to its not being equivalent to "pain is $R1 \vee R2$." It's hard to see
how all that could be so unless predicates with ellipses can refer.

To take for granted that (openly) disjunctive sentences (can) have truth
conditions is not, of course, to say what their truth conditions are. I haven't
got a semantics for such sentences to offer you; and trying to construct
one is not a task that I relish. Maybe they should be treated as true at
each world where any of the disjuncts is true, and as neither true nor false
(as noncommital, if you like) anywhere else. In what follows, I'll assume
some story of that sort; I doubt the details matter much for the polemical
purposes I have in mind.

Where we are now is: We're assuming that open disjunctions can
express bona fide properties, but that the properties that they express
are somehow *intrinsically* unfit for projection. This can really seem quite
puzzling. Presumably God can do anything He likes with the properties
He has at hand; so if there really are such properties as *being $R1 \vee R2 \ldots$*,
why can't God make laws about them? Nonetheless, I think the intuition
that open disjunctions are at best bad candidates for laws is basically
sound. Here's why: *Open laws suggest missed generalizations.* For, to offer a
law of the form $(R1 \vee R2 \vee \cdots) \rightarrow Q$ is to invite the charge that one has

failed correctly to identify the property in virtue of which the antecedent of the law necessitates the consequent. Or, to say the same thing the other way around: Someone who offers such a law undertakes a burden to provide a positive reason that there isn't a *higher level* but *nondisjunctive* property of things that are $R1 \lor R2 \ldots$ in virtue of which they bring it about that Q.

But we still haven't got to the bottom. No doubt, if there is a higher level property that subsumes all the states that satisfy an open disjunction, then we will want to formulate our laws in terms of it. But *why is that*? Not, surely, because we are prejudiced against disjunctions "as such"? Rather, I think, it's because formulas that express closed laws are *stronger* than the corresponding open ones, and ceteris paribus, we want to accept the strongest generalizations that our evidence confirms. Accepting the strongest generalizations that one's evidence confirms is *what induction is about*.

I'm suggesting that the intuition that open disjunctions invite higher level laws isn't different in kind from the intuition that open *lists* invite universal generalizations. *Swans are white* constrains more worlds (hence supports stronger counterfactuals) than *x is white if (x is swan a, ∨ x is swan b, ∨ x is swan c …)*. So too, and for the same reason, *pain causes avoidance* constrains more worlds than *x causes avoidance if (x is neural state a, ∨ x is neural state b …, ∨ x is silicon state f …, ∨ x is Martian state g, ∨ …)*. The cost of the former generalization is reifying not just the property of *being swan a or being swan b or being swan c*, etc. but also the more abstract, higher-level property of *being a swan*. The cost of the latter generalization is reifying not just the property of *being in neural state a or being in neural state b … or being in silicon state f … or being in Martian state g*, etc. but also the more abstract, higher-level property of *being in pain*. Pretty clearly, standard inductive practice is prepared to hypostasize in both cases. And the success of standard inductive practice suggests pretty clearly that it is right to do so.

It may be, according to this view of the matter, that there are laws about open disjunctive properties after all; God can do whatever He likes, as previously remarked. But at least it's apparent that we have general methodological grounds for preferring a nondisjunctive higher-level law to a corresponding lower-level disjunctive law, all else being equal. Induction is a kind of market prudence: Evidence is expensive, so we should use what we've got to buy all the worlds that we can.

Here's where the discussion stands now:

• Kim is wrong about what's wrong with jade. What's wrong with jade is not that it's MR but that it's a (closed) disjunctive property, and closed disjunctive properties are ipso facto not projectible.

• Kim is also wrong about the analogy between jade and pain. According to functionalists, pain, but not jade, is MR; that is, pain has an *openly* disjunctive realizer. It is to that extent okay for a functionalist to say both that there *aren't* laws about jade and that that there *are* laws about pain.

• But a functionalist who says this still needs to explain why we should (why we do) prefer higher-level nondisjunctive laws (*pain leads to avoidance*) to lower-level laws with open disjunctions (*states that are R1 ∨ R2 ∨ ... lead to avoidance*), all else being equal. Why are we prepared to buy nondisjunctive laws at the cost of reifying high-level properties? My story is that this policy complies with a dictate that inductive practice obeys quite generally: *Prefer the strongest claim compatible with the evidence, all else being equal*. Quantification over instances is one aspect of rational compliance with this injunction; reification of high level kinds is another.

So, then, everything is fine and all the mysteries—except, of course, the ones about induction itself—have dissolved? Not! (as I'm horrified to hear they say in California). I think, in fact, that what's really bugging Kim is indeed a metaphysical mystery about functionalism, and that the discussion we've been having so far hardly touches it. Let me try, in the closing section, to articulate what I take to be the trouble.

Conclusion (Molto Mysterioso)

Kim remarks, at one point, that "when we think about making projections over pain, very much the same worry should arise about their propriety as did for jade. Consider, a possible law: 'Sharp pains ... cause anxiety reactions.' Suppose this generalization has been well confirmed for humans. Should we expect on that basis that it will hold also for Martians whose psychology is implemented (we assume) by a vastly different mechanism? Not if we accept ... that psychological regularities hold, to the extent that they do, in virtue of the causal-nomological regularities at the physical implementation level" (324).

Apparently, Kim wants to get from premises 4 and 5 to the conclusion 6.

4. Psychological regularities hold only in virtue of implementation level regularities.

5. Martian pain is implemented by "vastly different" mechanisms than ours.

6. We shouldn't expect "pain" to be projectible over a mixed population of us and Martians; that is, we shouldn't expect that Martian pain will be like our pain in respects that are thus far unexamined.

But, on second thought, an enthymeme would seem to have crept into this line of argument. To warrant his inference, Kim also needs some such premise as 7.

> 7. The behaviors of systems that are "vastly different" at the physical level should be expected not to be reliably similar in respect of their nonphysical (e.g., of their higher-level or functional) properties. Such similarities as one finds are accidents (cf. the color of jadeA = the color of jadeB).

But, in respect to (7), one has serious reason to hesitate. For one thing, artifacts appear to offer an indefinite number and variety of counterexamples; that's why references to can openers, mousetraps, camshafts, calculators, and the like bestrew the pages of functionalist philosophy of science. To make a better mousetrap *is* to devise a new kind of mechanism whose behavior is reliable with respect to the high-level regularity "live mouse in, dead mouse out." Ceteris paribus, a *better* mousetrap is a mechanism that is *even more* reliable with respect to this regularity than mousetraps used to be. If it weren't possible, at least *sometimes*, for quite heterogeneous mechanisms to be reliable in respect of the functional descriptions that they converge upon, new kinds of mousetraps would never work; ingenuity would fail, and mousetrap makers would be doomed to careers of fruitless self-quotation. It looks, in short, like (7) might do as a rule of thumb, but it can't be anything like a natural necessity.

Likewise outside the realm of artifacts. The very *existence* of the special sciences testifies to reliable macrolevel regularities that are realized by mechanisms whose physical substance is quite typically heterogeneous. Does anybody really doubt that mountains are made of all sorts of stuff? Does anybody really think that, since they are, generalizations about mountains as such won't continue to serve geology in good stead? Damn near everything we know about the world suggests that unimaginably complicated to-ings and fro-ings of bits and pieces at the extreme *microlevel* manage somehow to converge on stable *macrolevel* properties. On the other hand, this "somehow" really is entirely mysterious, and my guess is that that is what is bugging Kim. He just doesn't see why there should be (how there could be) macrolevel regularities *at all* in a world where, by common consent, macrolevel stabilities have to supervene on a buzzing, blooming confusion of microlevel interactions. Or rather, he doesn't see why there should be (how there could be) unless, at a minimum, macrolevel kinds are *homogeneous* in respect to their microlevel constitution.[5] Which, however, functionalists in psychology, biology, geology, and elsewhere, keep claiming is typically not the case.

So, then, *why is there anything except physics*? That, I think, is what is *really* bugging Kim.[6] Well, I admit that I don't know why. I don't even know how to *think about* why. I expect to figure out why there is anything except physics the day before I figure out why there is anything at all, another (and, presumably, related) metaphysical conundrum that I find perplexing. I admit, too, that it's embarassing for a professional philosopher—a paid up member of the APA, Eastern Division, I assure you—to know as little as I do about why there is macrostructural regularity instead of just physical regularity. I would therefore quite like to take Kim's way out and dissolve the mystery by denying the premise. Kim wants to *just stipulate* that the only kinds there are (what he calls) *local*; that is, the only kinds there are the kinds of kinds whose realizers are physically homogeneous. "[T]he present view doesn't take away species-restricted mental properties ... although it takes away pain 'as such'" (25). More precisely, it doesn't take away species-specific mental properties unless it turns out that they are MR too; if they are, then of course it does.

In effect, Kim wants to make it true *by fiat* that the only projectible kinds are physically homogeneous ones. That's tempting, to be sure; for then the question why there are *macrolevel* regularities gets exactly the same answer as the question why there are *microlevel* regularities: *Both follow from physical laws*, where a physical law is, by definition, a law that has physical kinds on both ends. But, for better or worse, you don't get to decide this sort of thing by fiat; just as you don't get to avoid the puzzle about why there's something instead of nothing by stipulating that there isn't. Only God gets to decide whether there is anything, and, likewise, only God gets to decide whether there are laws about pains; or whether, if there are, the pains that the laws are about are MR. Kim's picture seems to be of the philosopher impartially weighing the rival claims of empirical generality and ontological transparency, and serenely deciding in favor of the latter. *But that picture won't do.* Here, for once, metaphysics actually matters,[7] so philosophers don't get to choose.

Science postulates the kinds that it needs in order to formulate the most powerful generalizations that its evidence will support. If you want to attack the kinds, you have to attack the generalizations. If you want to attack the generalizations, you have to attack the evidence that confirms them. If you want to attack the evidence that confirms them, you have to show that the predictions that the generalizations entail don't come out true. If you want to show that the predictions that the generalizations entail don't come out true, you have actually to *do the science*. Merely complaining that the generalizations that the evidence supports imply a philosophically inconvenient taxonomy of kinds *cuts no ice at all*. So far, anyhow, when the guys in the laboratories actually do the science, they keep finding that mental kinds are typically MR, but that the predictions that intentional psychology entails are, all the same, quite frequently confirmed. Lots of different sorts of microinteractions manage, somehow or other, to converge on much the same macrostabilities. The world, it seems, runs in parallel, at many levels of description. You may find that perplexing; you certainly aren't obliged to like it. But I do think we had all better learn to live with it.

Acknowledgment

Many thanks to Ned Block, Ernie Lepore, Joe Levine, Howard Stein, and especially Barry Loewer for helpful criticism of an earlier draft.

Notes

1. From here on, I'll honor the distinction between psychological (etc.) states and psychological (etc.) terms only where it matters. For much of the discussion we're about to have, it doesn't. Roughly, psychological states are what the terms in psychological theories denote if the theories are true.
2. "Real" clearly can't mean exceptionless in this context. But what Kim has against psychological laws isn't their failure to be strict. (Here again, contrast Davidson, 1980; see also Fodor, 1991 and Schiffer, 1991.)
3. More precisely: If jade is nephrite or jadeite, then it's metaphysically necessary that jade is nephrite or jadeite. I don't want to speculate on what we would (should) do if, for example, we were to find that there is yet a third kind of stuff in our jade samples.
4. I'm neutral on the hard question of whether a property can have a metaphysically possible realizer that isn't one of its nomologically possible realizers.
5. Note that what's mysterious isn't macrostructure per se; it's irreducible (autonomous) macrostructure. When macrokinds are metaphysically identical to microkinds, laws about the latter imply laws about the former; likewise when macroregularities are logical or mathematical constructions out of microregularities, as in the "Game of Life" described in Dennett, 1991. Pace Dennett, such cases do not illuminate (what functionalists take to be) the metaphysical situation in the special sciences. To repeat: autonomy implies "real" (viz., projectible) patterns *without reduction.*
6. To be sure, Kim is also bugged about problems of causal and explanatory overdetermination (see Kim, 1993, passim). But these are plausibly just the other side of the metaphysical problem about levels. Microlevel properties are projectible by consensus. If macrolevel properties can be both projectible and autonomous, it looks like a given causal transaction could be irreducibly "covered" by more than one causal law; and, presumably, reference to any of these covering laws could constitute a causal explanation of the transaction. That, I suppose, is what the issues about causal and explanatory overdetermination amount to.
7. And not just for psychology; the parallels in the current evolutionary wars are positively eerie. "[T]he ultra-Darwinians reveal a thoroughgoing reductionst stance. [They] simply wave at large-scale systems, but only address the dynamics of gene-frequency changes as they see them, arising from competitive reproductive struggle.... Naturalists, in contrast, are attuned to the hierarchical structure of biological systems. They are convinced that

there are processes relevant to understanding evolution that go on within each of these levels—from genes, right on up through populations, species, and ecosystems" (5). "The implacable stablity of species in the face of all that genetic ferment is a marvelous demonstration that large-scale systems exhibit behaviors that do not mirror exactly the events and processes taking place among their parts.... Events and processes acting at any one level cannot possibly explain all phenomena at higher levels (175). (Both quotes are from Eldredge, 1995. A super book, by the way, which everybody ought to read.)

Part II
Concepts

Chapter 3

Review of Christopher Peacocke's *A Study of Concepts*

The *Modern* era, as analytic philosophers reckon, started with Descartes. By contrast, the *Recent* era started when philosophy took the "linguistic turn" (Richard Rorty's phrase), hence with Frege or Russell, or early Wittgenstein, or the Vienna Circle—take your pick. Modern philosophy was mostly about epistemology; it wanted to understand what makes knowledge possible. Recent philosophy is mostly about meaning (or "content") and wants to understand what makes thought and language possible. So, anyhow, we tell our undergraduates when we're in a hurry.

There's something to it, but probably not much. "Transcendental" arguments used to run: "If it weren't that *P*, we couldn't *know* that *Q*; and we do know that *Q*; therefore *P*." Philosophical fashion now prefers: "If it weren't that *P*, we couldn't *say* (or think or judge) that *Q*; but we do say (or think or judge) that *Q*; therefore *P*." Much of a muchness, really. The two kinds of arguments tend to be about equally unconvincing and for the same reasons. Often enough, empiricist preconceptions are doing the work in both.

This is not, however, to deny that there is something very peculiar about Recent philosophy. There has indeed been a change, and it goes much deeper than shifting styles of philosophical analysis. What's really happened, not just in philosophy but in psychology, lexicography, linguistics, artificial intelligence, literary theory, and just about everywhere else where meaning and content are the names of the game, is a new consensus about what concepts are. Take a sample of current and Recent theorists, chosen with an eye to having as little else in common as may be: Heidegger, or Wittgenstein, or Chomsky, or Piaget, or Saussure, or Dewey, or any cognitive scientist you like, to say nothing of such contemporary philosophers as Davidson, Dennett, Rorty, and Quine. You may choose practically at random, but they are all likely to agree on this: *concepts are capacities*; in particular, concepts are *epistemic* capacities.

Christopher Peacocke's *A Study of Concepts* is about as subtle and sophisticated an elaboration of the idea that concepts are epistemic capacities as you will ever want to read. It may, in fact, be a *more* subtle and sophisticated elaboration than you will ever want to read. Peacocke is hard work and he spares his reader nothing. His prose is not, perhaps, denser than the intricacy of his thought requires, so I'm warning, not complaining; but his book wants exegesis, and it will surely get a lot. Many's the graduate seminar that will slog its way through, line by line, and will be edified by doing so.

I won't attempt anything of that sort here. There are too many passages that I do not understand, and of the ones I do understand, there are too many that I haven't made up my mind about. It does seem to me, however, that a striking number of Peacocke's moves depend upon assumptions that he makes, explicitly but practically without argument, in the book's first several pages. I propose to concentrate on these since, as I remarked, they strike me as a symptom of our times.

Peacocke's topic is the nature of concepts. Just roughly and by way of orientation:

> 1. *Concepts are word meanings.*[1] The concept DOG is what the word "dog" and its synonyms and translations express. This ties theories of concepts to theories of language.
> 2. *Concepts are constituents of thoughts.* To think that dogs bark is to entertain the concept DOG and the concept BARK.
> 3. *Concepts apply to things in the world.* The concept DOG is one which, of necessity, all and only dogs fall under. Judgments are applications of concepts, which is why it's things in the world that make judgments true or false.

This catalogue is mine, not Peacocke's, but I don't expect it's anything that he'd object to very much. So then, if that's what concepts are, what should a theory of concepts be?

Starting on page 5: "Throughout this book I will try to respect the following principle.... There can be nothing more to the nature of a concept than is determined by ... a correct account of 'grasping the concept'.... [A] theory of concepts should be a theory of concept possession." There are, to be sure, trivializing readings of this equation (C is the unique concept whose possession condition is that you have the concept C). But Peacocke intends that the nature of a concept should be illuminated by what a theory says about grasping it. For example (6), "Conjunction is that concept C to possess which a thinker must find [inferences of certain specified forms] primitively compelling, and must do so because they are of these forms." For example, it partially identifies C as the concept of conjunction that anybody who has it finds inferences from the premises p and q to the conclusion pCq primitively compelling as such.

Peacocke is saying that sometimes part of what grasping a concept comes to is being able to see, straight off, that certain of the inferences

that it figures in are okay. Another part of what grasping a concept sometimes comes to is being able to see straight off that the concept applies
to something one perceives. (This figures largely in the book, but I'll scant
it here. See the next two chapters in this volume.)

In either case, though Peacocke's main topic is the nature of conceptual
and/or linguistic content and is thus nominally about semantics, it's epistemology that is actually calling the shots. According to Peacocke, what
concepts you have depends on what concepts you have grasped, and
what concepts you have grasped depends on what *epistemic* capacities you
have acquired. Having the concept of CONJUNCTION, for example, is
being able to see certain inferences as valid. My point is that the putative linguistic turn, from a kind of philosophy that worries mostly about knowledge to a kind of philosophy that worries mostly about meaning, doesn't
actually amount to much if meaning is itself epistemically construed.

These days, the ideas that theories of concepts are theories of concept
possession and that possessing a concept is having certain epistemic
capacities are often treated as truisms. In fact, they are intensely tendentious. I suppose Descartes, or Hume, or Mill would have thought that you
identify a concept not by saying *what it is to grasp it* but by saying *what it
is the concept of*. Accordingly, on this older view, to illuminate the nature
of a concept you need not a theory of concept possession but a theory of
representation. The key question about, as it might be, the concept DOG
is not "what is it to have it," but something like "In virtue of what does
that concept represent dogs, and in virtue of what do other concepts fail
to do so?"

Changing the topic from "How do concepts represent?" to "What
capacities constitute the possession of a concept?" was, I think, what *really*
started Recent philosophy. And arguably Dewey and the Pragmatists had
more to do with it than Frege or Wittgenstein. It's a paradigmatically
Pragmatist idea that having a concept consists in being able to *do* something. By contrast, uninstructed intuition suggests that having a concept
consist in being able to *think* something. (Having the concept DOG is
being able to think about dogs; or, better, about the property of being a
dog.) In my view, it's uninstructed intuition that has this stick by the right
end.

So that joins the issue: Is a theory of concepts a theory of concept possession or is it a theory of how concepts represent? "Why not both?" you
might ask, in one of your ecumenical moods. But this would be to miss
the metaphysical tone of Peacocke's inquiry. No doubt there might be
both a theory of what it is to grasp a concept and a theory of how a concept represents. And there might also be theories of how a concept is
acquired and how it's applied, while we're at it. But, as Peacocke understands things, none of these would be a conceptual *analysis*—none of

them would count as philosophy, strictly speaking—unless it specifies the properties that make a concept the very concept that it is. Peacocke is saying that what makes a concept the very concept that it is are the conditions for possessing it.

Because he thinks philosophical analyses unpack the possession conditions that individuate concepts, Peacocke is prepared to take quite a hard line on the methodological priority of philosophical investigations in the cognitive sciences.

> An agenda for psychology suggested by the general approach I have been advocating is, then, this: For each type of thinker and for each concept possessed by a thinker of that type, to provide a subpersonal explanation of why the thinker meets the possession condition for that concept.... Carrying out this agenda is also in its very nature an interdisciplinary enterprise. For any particular concept, the task for the psychologist is not fully formulated until the philosopher has supplied an adequate possession condition for it. (190)

The idea that philosophy sets the agenda for psychology, or for any other empirical inquiry, is a typical product of the idea that philosophy is conceptual analysis and that conceptual analysis is, as Peacocke likes to say, "relatively a priori." It strikes me, frankly, as ahistorical and maybe a touch hubristic. The susurration that you hear is legions of cognitive psychologists not holding their breath until their task is fully formulated by philosophers.

I doubt that the theory of concepts that engenders this account of conceptual analysis can be sustained. The key problem is that people who have the concept DOG thereby have *all sorts* of capacities that people who don't have that concept thereby fail to have. And, surely, not all these capacities are essential to "the nature of the concept." If I didn't have the concept DOG, I suppose I couldn't have the concepts DOG BATH or DOG BONE or FIGHTING LIKE CATS AND DOGS. And, if I didn't have those concepts, there would be all sorts of inferences that I would fail to find "primitively compelling" (from "dog bath" to "bath for dogs," for example) and all sorts of perceptual judgments that I would be unable to make (that the object currently on display is a dog bath rather than a bird bath, for example). But I suppose (and so, I'm sure, would Peacocke) that none of *these* capacities illuminates the essential nature of the concept DOG; one could have the concept even if one had none of them.

If there is a capacity the possession of which is constitutive of grasping the concept DOG, and if having this capacity consists, inter alia, in being able to see certain inferences as primitively compelling, then there must be something that distinguishes such concept-constitutive inferences from the rest. And, as we've seen, it has to be something nontrivial if conceptual analysis is to be worth doing. (It's constitutive but trivial that DOG is a concept that lets you make inferences about dogs. It's nontrivial, but also nonconstitutive, that DOG is a concept that lets you make inferences about dog baths.) Are there some inferences that are both nontrivial and constitutive of the concept DOG? The answer *must* be "yes" if theories of concepts are theories of their possession conditions and possession conditions are inferential capacities.

Very well then, what determines which inferential capacities are nontrivially constitutive of grasping the concept *C*? Which of the cluster of capacities that grasping *C* may bring in train are the ones that belong to its possession conditions? Peacocke says a lot about which possession conditions are constitutive of one or other concepts, but remarkably little about the general question.

The closest we get is this (2): "Concepts *C* and *D* are distinct if and only if there are two complete propositional contents that differ at most in that one contains *C* substituted in one or more places for *D*, and one of which is potentially informative while the other is not." So, for example, the concept DOG is distinct from the concept BARKER because someone who has fully grasped the former concept, and who takes it that dogs are animals, might nevertheless take it to be news that barkers are animals. Whereas (assuming that DOG is the same concept as DOMESTIC CANINE) nobody who takes it that dogs are animals could find it news that domestic canines are too. Or, to put the same idea in terms of possession conditions: finding the inference *if dogs are animals then domestic canines are animals* primitively compelling is among the possession conditions for DOG. Whereas, though the inference *if dogs are animals then barkers are animals* is quite a good inference, finding it primitively compelling would presumably not count as constitutive for any of the concepts involved. This would be true even if, in point of fact, all and only dogs bark. (I emphasize that the example is mine and not Peacocke's, and that it is crude.)

The notion of an informative proposition (or inference) thus looms very large in Peacocke's treatment. He needs it a lot if he is to avoid the trivialization of his project. As far as I can tell, Peacocke thinks it's too obvious to bother arguing for that you can individuate possession conditions, and thereby flesh out the notion of concept identity, by appealing to the informativeness test. But I don't at all share his optimism.

Someone who finds it unsurprising that John understands that bachelors are bachelors might, I suppose, still wonder whether John understands that bachelors are unmarried men. So it appears that if, following Peacocke's recipe, you substitute "unmarried men" for the second "bachelors" in "John understands that bachelors are bachelors" you go from something unsurprising to something that someone might well take to be news. Since, however, the concepts BACHELOR and UNMARRIED MAN are identical if any concepts are, it looks like the informativeness test for concept identity is badly undermined. These so-called Mates cases (after their inventor, Benson Mates) are a philosophical commonplace; Peacocke doesn't mention them, but I don't understand why they don't worry him.

Or consider poor Jones, who went off with a bang; he didn't know that *being flammable* and *being inflammable* are the same thing. Jones would have found it informative had someone taken the trouble to tell him; if someone had, he would be with us still. Yet "flammable" and "inflammable" are *synonyms* and hence must express the same concept if concepts are word meanings. It looks, again, as though informativeness is one thing, conceptual identity another. (Peacocke has a remark on p. 32, arising in a quite different context, that may be intended to cover this sort of case: "In this particular example, it suffices for a theory of concepts to aim to explain those patterns of epistemic possibility that exist only for one who fully understands [the corresponding word] (*and any synonyms he may acquire*)" [Peacocke's parentheses, my emphasis]. I doubt, however, that Peacocke intends this as a codicil to his informativeness test, since it presupposes a notion of synonymy which is itself semantic and, to put it mildly, unexplicated.)

I am, truly, not meaning to quibble or to insist upon what are arguably marginal counterexamples. But, like lots of other philosophers who have been influenced by Quine, I really do doubt that concept identity can be explicated un-question-beggingly by appeal to notions like informativeness. I doubt, in fact, that it can be un-question-beggingly explicated *at all* so long as you think of having a concept in terms of possessing diagnostic *epistemic* capacities. (Whether concept identity can be explicated un-question-beggingly in *nonepistemic* terms is a long question to which the short answer is "maybe.") My own guess is that there aren't *any* nontrivial inferences that the concept DOG requires its possessors to have come what may. As Hilary Putnam has pointed out, even *dogs are animals* would fail in science-fiction worlds where dogs turn out to be Martian robots. In such a world, somebody could "fully grasp" the concept DOG but find the inference *dog → animal* uncompelling; indeed, unsound. Nor, I think, are there any perceptual judgments that DOG owners as such *must* be compelled to make. No landscape is so uncluttered that it is impossible *in principle* that one should fail to recognize that it contains a dog.

The problem is that Peacocke's whole project, his whole conception of what concepts are, and hence of what a theory of concepts should aim for, is committed to an epistemic distinction between analytic (constitutive) capacities and synthetic (collateral) capacities. And this is a distinction that Peacocke, like the rest of us, doesn't know how to draw. (He takes note of this commitment in a passing footnote on p. 243 and is unperturbed by it.) This leads to a geographical impasse: If, as people on my side of the Atlantic are increasingly inclined to suppose, there isn't an epistemic analytic/synthetic distinction, then the notion of a possession condition is infirm and you can't identify grasping a concept with being disposed to draw the inferences by which its possession conditions are constituted. Correspondingly, the philosophical analysis of a concept can't consist in setting out the possession conditions of the concept.

The long and short is: I think there is good reason to doubt that the kind of philosophy Peacocke wants to do can be done. In one passage, Peacocke remarks almost plaintively that "theories are developing in the literature of what it is to possess certain specific concepts.... While there is much that is still not understood and not all of what has been said is right, it is hard to accept that the goal of this work is completely misconceived" (35–36). I guess I don't find it all that hard. The linguistic turn was, I think, an uncompleted revolution; to *really* turn from theories of knowledge to theories of meaning, you would have to stop construing content in epistemological terms. Many analytic philosophers can't bear not to construe content in epistemological terms because they think of philosophy as conceptual analysis, and of conceptual analysis as displaying a concept's possession conditions, and of possession conditions as characteristically epistemic.[2] If, as I believe, that whole picture is wrong, a certain kind of analytic philosophy is ripe for going out of business. *If there is no analytic/synthetic distinction, then there are no analyses.* This is a thought that keeps philosophers on my side of the Atlantic awake at night. Why doesn't it worry more philosophers on Peacocke's side?

Does any of this really matter except to philosophers over sherry? Oddly enough, I think perhaps it does. We are in the midst of a major interdisciplinary attempt to understand the mental process by which human behavior accommodates to the world's demands—an attempt to understand human rationality, in short. Concepts are the pivot that this project turns on since they are what mediate between the mind and the world. Concepts connect with the world by representing it, and they connect with the mind by being the constituents of beliefs. If you get it wrong about what concepts are, almost certainly you will get the rest wrong too.

The cognitive scientists I know are mostly a rowdy and irreverent lot, and I shouldn't want to be around when they hear Peacocke's views about the primacy of philosophy in defining their enterprise. But it's perfectly true that they have, almost without exception, assumed what is essentially a philosophical theory of concepts, and that it's pretty much the one that

Peacocke also takes for granted: concepts are epistemic capacities. In consequence, questions about which epistemic capacities constitute which concepts perplex the whole discipline, and nobody knows any more than Peacocke does how to answer them. If it turns out that concepts *aren't* epistemic capacities, these questions *don't have* answers.

I'm not proposing a trans-Atlantic methodological shoot-out, but I do think there needs to be a sustained discussion of what concepts are, and I think that de-espistemologizing semantics—completing the linguistic turn—is likely to be its outcome. If so, theories of language and mind will eventually come to look very different from what Recent philosophy has supposed; and the project of philosophical analysis will come to look inconceivably different. (Assuming, indeed, that there is any such project left.) In this discussion, someone will have to speak with insight and authority for the Received View. Peacocke has done that, and we are all in his debt.

Notes

1. Which is not to say that word meanings are ipso facto concepts; I suppose Peacocke would deny that they are in the case of demonstratives or names, for example. Nor does he claim (nor should he) that for each of our concepts we must have a corresponding word with which to express it.

   Getting clear on the word-concept relation is no small matter. Suffice it, for purposes of the present discussion, that Peacocke thinks that at least some words mean what they do because they express the concepts that they do; and I think he's right.

2. It's not, of course, unintelligible that analytic philosophers should have assumed all this. Transcendental arguments are supposed to ground antiskeptical conclusions. The plan is to say to the skeptic something like the following: "Look, if you don't recognize even *this* as a dog (or, mutatis mutandis, "if you don't accept even that dogs are animals") then you simply haven't got the concept. I can only argue about dogs with someone who *does* have the concept." But, of course, that kind of line won't work unless the connection between the concept and the corresponding epistemic capacities is constitutive.

   Lots of philosophers fear that if concepts don't have analyses, justification breaks down. My own guess is that concepts *don't* have analyses and that justification will survive all the same.

Chapter 4

There Are No Recognitional Concepts—Not Even RED

Introduction

Let it be that a concept is *recognitional* if and only if:

1. It is at least partially constituted by its possession conditions; and
2. Among its possession conditions is the ability to recognize at least some things that fall under the concept *as* things that fall under the concept.

For example, RED is a recognitional concept iff it numbers, among its constitutive possession conditions, the ability to recognize at least some red things as red.

In this paper, I propose to argue—indeed, I propose to sort of *prove*—that there are no recognitional concepts; not even RED.

Lots of philosophers are sympathetic to the claim that there are recognitional concepts. For one thing, insofar as *recognitional* capacities are construed as *perceptual* capacities, the claim that there are recognitional concepts preserves the basic idea of Empiricism: that the content of at least some concepts is constituted, at least in part, by their connections to percepts. For philosophers who suppose that Empiricism can't be *all* wrong, recognitional concepts can therefore seem quite a good place to dig in the heels. More generally, the claim that there are recognitional concepts is a bastion of last resort for philosophers who think that semantic facts are constituted by epistemological facts, a doctrine that includes, but is not exhausted by, the various forms of Empiricism. If you have ever, even in the privacy of your own home among consenting adults, whispered, hopefully, the word "criterion," then probably *even you* think there are recognitional concepts.

Philosophers who hold that there are recognitional concepts generally hold that it's important that there are; for example, a familiar line of anti-skeptical argument turns on there being some. The idea is that, if a concept is recognitional, then having certain kinds of experience would, in principle, *show with the force of conceptual necessity* that the concept applies. If, for example, RED is a recognitional concept, then having certain kinds of experience would, in principle, show with the force of conceptual necessity that there are red things. Ditto, mutatis mutandis, SQUARE, CHAIR, IS IN PAIN, and BELIEVES THAT P, assuming that these are recognitional concepts. So, if you think that it's important that skepticism about squares, chairs, pains, beliefs, or red things be refuted, you are likely to want it a lot that the corresponding concepts are recognitional. Nevertheless, it's sort of provable there aren't any recognitional concepts; so, at least, it seems to me.

I pause to mention a kind of argument against there being recognitional concepts to which I am sympathetic, but which I am *not* going to pursue in what follows: namely, that it's truistic that the content of one's experience underdetermines the content of one's beliefs, excepting only one's beliefs about one's experiences. No landscape is so empty, or so well lit— so the thought goes—that your failure to recognize that it contains a rabbit *entails* that you haven't got the concept RABBIT. So, it couldn't be that your having the concept RABBIT requires that there are circumstances in which you couldn't but recognize a rabbit as such.

I think this is a good argument, but, notoriously, lots of philosophers don't agree; they think, perhaps, that the connection between concept possession and recognitional capacities can be relaxed enough to accommodate the truisms about rabbits without the claim that there are recognitional concepts lapsing into vacuity. I propose, in any event, not to rely upon this sort of argument here.

Compositionality

The considerations I will appeal to are actually quite robust, so a minimum of apparatus is required to introduce them. It will, however, be useful to have on hand the notion of a *satisfier* for a concept. The satisfier(s) for a concept are the states, capacities, dispositions, etc. in virtue of which one meets the possession condition(s) for the concept.[1] So, if the ability to tell red from green is a possession condition for the concept RED, then *being able to tell red from green* is a satisfier for the concept RED. If a disposition to infer P from P & Q is a possession condition for the concept conjunction, then *being so disposed* is a satisfier for the concept conjunction. And so forth. Since, by assumption, concepts have their possession conditions essentially, and possession conditions have their satisfiers essentially, the exposition will move back and forth between the three as convenience dictates.

I propose to argue that there are no concepts among whose satisfiers are recognitional capacities, hence that there are no recognitonal concepts. I need a premise. Here's one:

Premise P: S is a satisfier for concept C if and only if C inherits S from the satisfiers for its constituent concepts.[2] (I'll sometimes call this the "compositionality condition" on concept constitution.)

Consonant with my general intention not to have the argument turn on its details, I leave it open how "inherited from" is construed in premise *P*, so long as fixing the satisfiers for constituent concepts is necessary and sufficient for fixing the satisfiers for their hosts. I do, however, urgently call your attention to the following point, ignoring which breeds monsters. Suppose concept C is a constituent of concept H(ost). Suppose, also, that S is a satisfier for C. Then, of course, perfectly trivially, since you can't have H unless you have C, you can't have H unless you have a concept one of whose satisfiers is S. What does *not* follow, however, is that you can't have H unless you have S. To get that conclusion you need the further principle that you can't have a concept unless you have the satisfiers for its constituents; that is, the principle that hosts "inherit" the satisfiers of their constituents. This is untrivial, and it is what premise P asserts.

Why Premise P is Plausible
Unless P is true, we will have to give up the usual account of why concepts are systematic and productive; and, mutatis mutandis, of how it is possible to learn a language by learning its finite basis. Consider, for example, the concept-constitutive possession conditions for the concept RED APPLE. If premise *P* is false, the following situation is possible: The possession conditions for RED are *ABC* and the possession conditions for RED APPLE are *ABEFG*. So denying P leaves it open that one could have the concept RED APPLE and not have the concept RED.

But, now, the usual compositional account of productivity requires that one satisfy the possession conditions for complex concepts, like RED APPLE, *by* satisfying the possession conditions for their constituent concepts. That is, it requires that one's having a grasp of the concept RED is *part of the explanation* of one's having a grasp of the concept RED APPLE. So accepting the usual compositional account of productivity is incompatible with denying premise P.

Likewise the other way around. The usual compositional account of productivity requires that if one satisfies the possession conditions for the constituents of a complex concept, one thereby satisfies the possession conditions for the concept.[3] But, suppose that premise P is false, and consider, once again, the concept RED APPLE. Denying P leaves it open that the concept-constitutive possession conditions for RED APPLE are *not* exhausted by the concept-constitutive possession conditions for RED and APPLE. For example, the former might be *ABCDE* and the latter might be *AB* and *CD* respectively. But then grasping the concepts RED and APPLE would not be sufficient for grasping the concept RED APPLE, and, once again, the standard account of conceptual productivity would be undermined.

So much for the bona fides of premise P. The next point is that the condition that compositionality imposes on concept constitution is highly substantive. A brief digression will show the kind of theory of concepts that it can rule out.

Consider the idea that concepts are (or are partially) constituted by their stereotypes, hence that *knowing its stereotype* is a satisfier for some concepts. Premise P says that this idea is true only if, if you know the stereotypes for the constituents of a complex concept, then you know the stereotype for that concept. Which, in some cases, is plausible enough. Good examples of RED APPLES are stereotypically red and stereotypically apples. Let's assume that that's because *the stereotype of RED APPLE is inherited from the stereotype for RED and the stereotype for APPLE.* (This assumption is concessive, and it may well not be true. But let it stand for the sake of the argument.) So then, as far as RED APPLE is concerned, it's compatible with premise P that knowing their stereotypes should be possession conditions for RED and APPLE.

Still, concepts can't be constituted by their stereotypes (knowing its stereotype can't be a satisfier for a concept). That's because RED APPLE isn't the general case. In the general case, complex concepts don't inherit their stereotypes from those of their constituents. So, in the general case, stereotypes don't satisfy premise P.

Consider such concepts as PET FISH, MALE NURSE, and the like.[4] You can't derive the PET FISH stereotype from the FISH stereotype and the PET stereotype. So, if stereotypes were constitutive of the corresponding concepts, having a grasp of FISH and having a grasp of PET (and knowing the semantics of the *AN* construction; see fn. 3) would *not* suffice for having a grasp of PET FISH. So, the usual story about how PET FISH is compositional would fail.

So much for stereotypes. If premise P is true, it follows that they can't be concept-constitutive. I will now argue that, if premise P is true, then it likewise follows that there are no recognitional concepts.

In fact, most of the work is already done, since for all intents and purposes, the notion of a recognitional concept is hostage to the notion that concepts are constituted by their stereotypes. Here's why. Nobody could (and nobody does) hold that the possession of a recognitional concept requires being able to identify *each* of its instances as such; if that *were* the requirement, then only God would have any recognitional concepts. So, the doctrine must be (and, as a matter of fact, always is) that possession of a recognitional concept requires the ability to identify good instances as

such in favorable conditions. (There are various variants of this in the literature; but it doesn't matter to what follows which you choose.)[5]

But now, unsurprisingly, the ability to recognize *good instances* of Fs doesn't compose, and this is for exactly the same reason that *knowing the stereotype of F* doesn't compose; good instances of F & Gs needn't be either good instances of F or good instances of G. See PET FISH, once again: Good instances of PET FISH are, by and large, poorish instances of PET and poorish instances of FISH. So a recognitional capacity for good instances of PET and good instances of FISH is not required for, and typically does not provide, a recognitional capacity for good (or, indeed, any) instances of PET FISH.

Somebody who is good at recognizing that trouts are fish and that puppies are pets *is not thereby good at recognizing that goldfish are pet fish*. The capacity for recognizing pet fish as such is not conceptually, or linguistically, or semantically connected to capacities for recognizing pets as such or fish as such. The connection is *at best* contingent, and it's entirely possible for any of these recognitional capacities to be in place without any of the others.

This doesn't, of course, show that the semantics of PET FISH are uncompositional. What it shows is that recognitional capacities aren't possession conditions for the concepts that have them. If recognitional capacities were possession conditions, PET FISH would not inherit its satisfiers from those of PET and FISH. So if recognitional capacities were possession conditions, PET FISH would fail premise P. So recognitional capacities aren't possession conditions. So there are no recognitional concepts.[6]

Objections

Q1: You're, in effect, taking for granted not only that compositionality is needed to explain productivity, but that it is therefore a test for whether a property is constitutive of the concepts that have it. Why should I grant that?

A1: I suppose I could just dig my heels in here. Compositionality is pretty nearly all that we know about the individuation of concepts. If we give that up, we will have no way of distinguishing what *constitutes* a concept from such of its merely contingent accretions as associations, stereotypes, and the like.

But though I do think it would be justifiable to take that strong line, I really don't need to in order to run the sort of argument I'm endorsing. If push comes completely to shove, the following exiguous version will do for my polemical purposes:

If you know what "pet" and "fish" mean, you thereby know what "pet fish" means. But you can be able to recognize pets and fish as such but be quite unable to recognize pet fish as such. So recognitional capacities can't be meanings, and they can't be constituents of meanings, all varieties of Empiricist semantics to the contrary notwithstanding.

As far as I can see, that formulation doesn't need much more than the distinctness of discernibles, so it seems to me that it cuts pretty close to the bone.

Q2: But couldn't an Empiricist just *stipulate* that recognitional capacities, though they demonstrably don't satisfy premise P and are thus demonstrably not constituents of meanings, are nevertheless to count as essential conditions for the possession of primitive concepts?

A2: Sure, go ahead, stipulate; and much joy may you have of it. But nothing is left of the usual reasons for supposing that the concepts we actually have comply with the stipulation; in particular, nothing is left of the idea that the *content* of our concepts is constituted by our recognitional capacities. Whatever content is, it's got to be compositional; so it's got to come out that the content of RED APPLE includes the content of RED and the content of PET FISH includes the content of PET.

It may be worthwhile to reiterate here an argument I gave for the plausibility of premise P. Suppose a primitive concept has a possession condition that is *not* inherited by one of its complex hosts; suppose, for example, that being able to recognize good instances of pets is a possession condition for PET but is *not* a possession condition for PET FISH. Then presumably it is possible that someone who has the concept PET FISH should nonetheless not have the concept PET. I take this to be a reductio, and I think that you should too.

Here's a closely related way to make the same argument: Perhaps you're the sort of philosopher who thinks it's a possession condition for RED APPLE that one is prepared to accept the inference RED APPLE → RED (i.e., that one finds this inference "primitively compelling").[7] If so, that's all the more reason for you to hold that the possession conditions for RED APPLE must *include* the possession conditions for RED. Hence it's all the more reason for you to hold that the satisfiers for RED are inherited under composition by RED APPLE. But if that's right, then once again it couldn't be that a recognitional capacity is a satisfier for RED *unless* it's a satisfier for RED APPLE. But the capacity to recognize pets as such is not a satisfier for the concept PET FISH, so it can't be a satisfier for PET. Since sauce for the goose is sauce for the gander, the ability to recognize red things is likewise not a satisfier for RED.

Q3: Couldn't we split the difference? Couldn't we say that the satisfiers for the *primitive* concepts include recognitional capacities, but that the satisfiers for complex concepts don't?

A3: Simply not credible. After all, people who have the concept PET FISH do generally have a corresponding recognitional capacity; for example, they are generally good at recognizing goldfish as pet fish. And, surely, being able to recognize (as it might be) a trout as a fish stands in *precisely* the same relation to having the concept FISH as being able to recognize a goldfish as a pet fish does to having the concept PET FISH. So, how could that relation be constitutive of concept possession in the one case but not in the other? Is it, perhaps, that the concepts FISH and PET FISH have content in different senses of "content"?

This sort of point is probably worth stressing. Some philosophers have a thing about recognitional capacities because they want to tie meaning and justification close together (see the remarks earlier about antiskeptical employments of the idea that there are recognitional concepts). But if recognitional capacities are constitutive only of primitive concepts, then the connection between meaning and justification fails in infinitely many cases. It will thus be concept-constitutive that (ceteris paribus) *it's trout-looking* is evidence for "it's a fish," but *not* concept constitutive that (ceteris paribus) *it's goldfishlooking* is evidence for "it's a pet fish." What possible epistemological use could such a notion of concept-constitutivity be put to?

Q4: FISH and PET are only *relatively* primitive (they're only primitive relative to PET FISH). What about absolutely primitive concepts like RED? Surely the concept RED is recognitional even if neither FISH nor PET FISH is.

A4: It's just more of the same. Consider RED HAIR, which, I will suppose, is compositional (that is, not idiomatic) and applies to *hair that is red as hair goes*. This view of its semantics explains why, though red hair is arguably not literally red, still somebody who has RED and has HAIR and who understands the semantic implications of the syntactic structure AN, can figure out what "red hair" means. So, prima facie, RED HAIR is compositional and the demands of productivity are satisfied according to the present analysis.[8]

But notice, once again, that the productivity/compositionality of the concepts does not imply the productivity/compositionality of the corresponding recognitional capacities. Somebody who is able to say whether something is a good instance of HAIR and whether something is a good instance of RED is not *thereby* able to recognize a good instance of RED HAIR. Well then, what *does* "red" contribute to the semantics of "red hair"? Just what you'd suppose: it contributes a reference to the property of *being red* (as such). It's just that its doing that isn't tantamount to, and doesn't entail, its contributing a recognitional capacity for (good instances of) redness.

One's recognitional capacity for RED doesn't compose. So one's recognitional capacity for red things is not a satisfier for the concept RED. So not even RED is a recognitional concept.

Q5:   What do you say about intentional concepts?

A5:   Nothing much for present purposes. They have to be compositional, because they are productive. If they are compositional, then there are, to my knowledge, three theories (exhaustive but not exclusive) of what they inherit from their constituents:

- they inherit the *extensions* of their constituents;
- they inherit the *senses* of their constituents;
- the inherit the *shapes* of their constituents. (Notice that shape is compositional; "red hair" contains "red" as a morphosyntactic part; and the shape of "red hair" is completely determined given the shape of "red," the shape of "hair," and the morphosyntactics of the expression.)

But, invariably a theory of intentional concepts that says that any of these are *inheritable* properties of their constituents will also say that they are *constitutive* properties of their constituents. So, as far as I can tell, nothing that anybody is likely to want to say about intentional concepts will deny my argument the premise it requires.

Conclusion

The moral of this chapter is that recognitional capacities are contingent adjuncts to concept possession, much like knowledge of stereotypes; a fortiori, they aren't *constitutive* of concept possession. How, indeed, *could* anyone have supposed that recognitional capacities are satisfiers for concepts, when recognitional capacities patently don't compose and concept satisfiers patently do?

I think what went wrong is, after all, not very deep, though it's well worth attending to. Content, concept-constitutivity, concept possession, and the like, are connected to the notion of an *instance* (i.e., to the notion of an *extension*). The notion of an instance (extension) is semantic, hence compositional, through and through; idioms excepted, what is an instance of a complex concept depends exhaustively on what are the instances of its parts. The notion of *a recognitional capacity*, by contrast, is connected to the notion of a *good* (in the sense of a typical, or an epistemically reliable) instance; the best that a recognitional capacity can promise is to identify good instances in favorable conditions. It's a mistake to try to construe the notion of an instance in terms of the notion of a good instance;[9] unsurprisingly, since the latter is patently a special case of the former, the right order of exposition is the other way around.

Recognitional capacities don't act like satisfiers: What's a satisfier for a complex concept depends on what's a satisfier for its parts; but what's a good instance of a complex concept doesn't depend on what's a good instance of its parts. Why should it? What's a good instance of a concept, simple or complex, depends on *how things are in the world*.[10] Compositionality can tell you that the instances of PET FISH are all and only the pet fish; but it can't tell you that the *good* instances of pet fish are the goldfish; which is, all the same, the information that pet fish *recognition* (as opposed to mere PET FISH *instantiation*) is likely to depend on. How could you expect semantics to know what kind of fish people keep for pets? Likewise, what counts as red hair depends, not just on matters of meaning, but also on what shades hair actually comes in (i.e., because red hair is hair that is *relatively* red.) How *could* you expect semantics to know what shades hair actually comes in? Do you think that semantics runs a barber shop?[11] How, in short, could you expect that relations between recognitional capacities would exhibit the compositionality that productivity requires of semantic relations?

Oh, well; so what if there are no recognitional concepts?

For one thing, as I remarked at the outset, if there are no recognitional concepts we lose a certain class of antiskeptical arguments; ones that depend on the connection between percepts and concepts being, in some sense, constitutive. We will no longer be able to say to the skeptic: "If you don't think that *this* experience shows that *that's* a chair, then you don't have the concept CHAIR." But maybe this isn't a great loss; I've never heard of a skeptic actually being convinced by that kind of argument. I sure wouldn't be if I were a skeptic.

I'm not, however, meaning to deny that the issue about recognitional concepts goes very deep. To the contrary, I'm meaning to claim that it goes very much deeper than (mere) epistemology. Close to the heart of the last hundred years of philosophy is an argument between a Cartesian and a Pragmatist account of concept possession. Though the details vary, the essentials don't: According to Cartesians, having the concept X is being able to *think about* Xs; according to Pragmatists, its being able to *respond differentially* or *selectively* to Xs (for short: it's being able to *sort* Xs.) I doubt that there's a major philosopher, anyhow since Peirce—and including, even, the likes of Heidegger—who hasn't practically everything at stake on how this argument turns out.

Notice that the issue here isn't "Naturalism." Sorting is just as intentional as thinking, and in the same way: *Neither* coextensive thoughts *nor* coextensive sorts are ipso facto identical. A Pragmatist who isn't a behaviorist can (and should) insist on this. The issue, rather, is whether the intentionality of thought derives from the intentionality of action. Roughly, Pragmatists think that it does, whereas Cartesians think that the metaphysical dependencies go the other way around. It's the difference between holding, on the one hand, that whether you are sorting Xs is a matter of how you are thinking about what you are doing; or, on the other hand, that whether you are thinking about Xs depends on (possibly counterfactual) subjunctives about how you would sort them.

Well, the minimal Pragmatist doctrine (so it seems to me) is the claim that there are recognitional concepts; that is, that at least some concepts are constituted by one's ability to sort their instances. And the present arguments (so it seems to me) show that even this minimal Pragmatist doctrine isn't true. Thinking centers on the notion of an instance; recognitional capacity centers on the notion of a *good* instance. Unless you are God, whether you can recognize an instance of X depends on whether it's a good instance of an X; the less good it is, the likelier you are to fail.[12]

But you can always *think* an instance of X; namely, by thinking *an instance of X*. So thinking is universal in a way that sorting is not. So thinking doesn't reduce to sorting. That is bedrock. To try to wiggle out of it, as so many philosophers have drearily done, by invoking ideal sorts, recognition under ideal circumstances, the eventual consensus of the scientific community, or the like, is tacitly to give up the defining Pragmatist project of construing semantics epistemologically. *Being an ideal sort* always turns out not to be independently definable; it's just being a sort that gets the extension right.

Or, to put the point with even greater vehemence: The question whether there are recognitional concepts is really the question what thought is *for*; whether it's for directing action or for discerning truth. And the answer is that Descartes was right: The goal of thought is to *understand* the world, not to sort it. That, I think, is the deepest thing that we know about the mind.

Afterword

This paper was presented at the 1997 meeting of the Central Division of the American Philosophical Association. Stephen Schiffer commented, and what he said was typical of the reaction I've had from a number of philosophical friends. So I include here my reply to Steve's reply.

Steve asked, in effect: "What's wrong with a mixed view, according to which recognitional capacities are constitutive for (some) primitive concepts but not for their complex hosts?" Steve thinks that my reply must be either aesthetic (mixed theories are ugly) or an outright appeal to the "agglomerative principle" that if the conjuncts of a conjunctive proposition are recognitional, then so too is the conjunctive proposition. Since Steve takes this principle to be not better than dubious, he thinks that I haven't a better than dubious argument against there being recognitional concepts.

Now, it does seem to me that somebody who holds that there are primitive recognitional concepts should also hold the agglomerative principle (see the discussion of Q3). But my thinking this isn't an essential part of my argument. I tried to make clear in the text what the essence of my argument is; but, evidently, I didn't succeed. This hasn't been my century for making things clear.

Here it is again:

A theory of compositionality should explain why, in the standard case, anybody who has a complex concept also has its constituent concepts (why anybody who has RED TRIANGLE has RED and TRIANGLE; why anybody who has GREEN HAIR has GREEN and HAIR ... and so forth). This is tantamount to saying that a compositionality principle should be so formulated as to entail that satisfying the possession conditions for a complex concept *includes* satisfying the possession conditions for its constituents.

Now look at Steve's proposal, which is that "it is reasonable to hold that the possession conditions of complex concepts are determined by those of their constituents concepts. But for the case at hand this simply requires F & G to be such that to possess it one must be able to recognize good instances of F and good instances of G."

As stated, Steve's theory is wrong about PET FISH: It's true, of course, that to have the concept PET FISH you have to have the concept FISH. But it's certainly *not* true that to have the concept PET FISH you have to have a recognitional capacity for good instances of fish. To have a concept, conjunctive or otherwise, you have to have the concepts that are its constituents. But you *don't* have to have recognitional capacities corresponding to its constituents; *not even if, by assumption, the complex concept is itself recognitional.* So, having the constituents of a concept can't require having a recognitional capacity in respect of their instances. If it did, you could have a complex concept *without* having its constituents—which is not an option. So concepts can't be recognitional capacities.

Can Steve's proposal be patched? Well, if he is to get the facts to come out right, he'll presumably just have to stipulate that for *some F & G* concepts (RED TRIANGLE) "the possessor must be able to recognize good instances of F and G," but that for others (PET FISH, MALE NURSE) that's not required. And he'll have to say, in some general and principled way, which concepts are which. But, surely, you don't want to have to stipulate the relations between the possession conditions for a complex concept and the possession conditions for its constituents; what you want is that they should just fall out of the theory of compositionality together with the theory of concept constitutivity.

Which, indeeed, they do, if you get these theories right. What's constitutive of FISH, and hence what PET FISH inherits from FISH, is (not a capacity for recognizing fish but) *the property that FISH expresses: namely, the property of being a fish.* Likewise, mutatis mutandis, what's constitutive of RED, and hence what RED TRIANGLE inherits from RED, is (not a recognitional capacity for red things but) the property that RED expresses, namely, the *property of being RED.* Given that what is constitutive of a concept determines its possession conditions, it follows that you can't have the concept PET FISH unless you know that pet fish are fish, and you can't have the concept RED TRIANGLE unless you know that red triangles are red. This is, of course, just what intuition demands.

Explanations are better than stipulations; and they're a *lot* better than stipulations that misdescribe the facts. So there *still* aren't any recognitional concepts.[13]

Acknowledgment

Thanks to Ned Block, Paul Horwich, Chris Peacocke, Stephen Schiffer, and Galen Strawson for helpful comments on an earlier draft of this essay.

Notes

1. Whereas, by contrast, the satisfiers *of* a concept are just whatever is in its extension. This is not, admittedly, a very happy way of talking, but it's no worse, surely, than intention/intension, and nothing better came to mind.
2. If *C* is a primitive concept, the condition is trivially satisfied.
3. This isn't quite right, of course; you also have to know how the constituents are "put together." Suppose a satisfier for RED is being able to identify red things and a satisfier for SQUARE is being able to identify square things. Then, presumably, the corresponding satisfier for RED SQUARE is being able to identify things in the intersection of RED and SQUARE. The fact that it's the intersection rather than, say, the union, that's at issue corresponds to the structural difference between the concept RED SQUARE and the concept RED OR SQUARE.
4. The immediately following arguments are familiar from the cognitive science literature on stereotypes, so I won't expand on them here. Suffice it to emphasize that the main point—that stereotypes don't compose—holds whether stereotypes are thought of as something like exemplars or as something like feature sets. For a review see Fodor and Lepore, 1992.
5. The intended doctrine is that having the recognitional concept *F* requires being able to recognize good instances of *F* as instances of *F*, not as *good* instances of *F*. It's concessive of me to insist on this distinction, because it requires the empiricist to defend only the weaker of the two views.
6. This assumes, of course, that what holds for PET and for FISH holds likewise for *any* candidate recognitional concept: namely, that there will always be *some* complex concept

of which it is a constituent but to which it does not contribute its putative possession condition. The reader who doubts this should try, as an exercise, to find a counterexample. For starters, try it with RED and APPLE.

7. Even conceptual atomists like me can hold that inferences that relate a complex concept to its parts are typically analytic and concept-constitutive. See Fodor and Lepore, 1992.

8. Let it be that an *AN* concept is "intersective" if its extension is the intersection of the *A*s with the *N*s. The standard view is that being intersective is sufficient but not nececcessary for an *AN* concept to be compositional: RED HAIR is *compositional* but not *intersective*, and PET FISH is both. (For a general discussion, see Kamp and Partee, 1995). Actually, my guess is that RED HAIR, BIG ANT, and the like are the general case. Excepting the "antifactive" adjectives ("fake," "imitation." etc.), *AN* usually means *A FOR (an) N*, and the intersectives are just the limiting case where things that are *A* for (an) *N* are *A*.
    But it doesn't matter for present purposes whether this is so.

9. Indeed, it's a venerable mistake. I suppose the Platonic theory of Forms was the first to commit it.

10. It also depends on how things are with us. What the good instances of RED are almost certainly has to do with the way the physiology of our sensory systems is organized (see, for example, Rosch, 1973; Berlin and Kay, 1969). Likewise, it's no accident that the good instances of ANIMAL are all big enough for us to see (i.e., big enough for *us* to see). It does not follow that a creature whose range of visual acuity is very different from ours would *thereby* have a different concept of animals from ours.

11. The cases in the text are not, of course, exceptional. What counts as an average income depends not only on what "average" and "income" mean, but also on what incomes people actually earn. Semantics tells you that the average income is in the middle of the income distribution, whatever the distribution may be. But if you want to *recognize* an average income, you need the facts about how many people actually earn how much. Semantics doesn't supply such facts; only the world can.

12. Analogous remarks hold for other epistemological capacities like, e.g., drawing inferences. Unless you are God, whether in a particular case you are disposed to infer *P* from *P* & *Q* depends, inter alia, on whether the logical form of the proposition is perspicuous. This strongly suggests that the considerations that rule out recognitional capacities as concept-constitutive will apply, mutatis mutandis, to rule out *any* epistemological candidate.

13. Steve also suggested that maybe PAIN is a recognitional concept, even if RED is not. I won't, however, discuss the notion that sensation concepts might be recognitional since I guess I don't really understand it. Does one recognize one's pains when one has them? Or does one just have them? If I can indeed recognize good instances of MY PAIN, I suppose it follows that I have the concept PAIN. Does it *follow*, as compositionality would require if PAIN is a recognition concept, that I can also recognize good instances of YOUR PAIN?
    Hard cases make bad laws. Sensation concepts are too hard for me.

Chapter 5

There Are No Recognitional Concepts—Not Even RED, Part 2: The Plot Thickens

---

Introduction: The Story 'til Now

Some of the nastiest problems in philosophy and cognitive science are either versions of, or live nearby, what I'll call question Q:

> Q: What are the essential (constitutive) properties of a linguistic expression *qua* linguistic?

Here are some currently live issues to which I suppose (and to which I suppose I suppose untendentiously) an answer to Q would provide the key:

- What do you have to learn (know, master) to learn (know, master) a linguistic expression (concept)? Variant: What are the "possession conditions" for a linguistic expression (concept)?[1]
- What is the principle of individuation for linguistic expressions?
- What makes two linguistic tokens tokens of the same linguistic type?
- Suppose G is the grammar of language L and E is a *lexical expression* in L (roughly, a word or morpheme). What sort of information about E should G contain?
- What's the difference between linguistic and "encyclopaedic" knowledge?
- What belongs to the "mental representation" of a linguistic expression as opposed to the mental representation of its denotation?
- Assume that some of the inferences that involve a lexical item are constitutive. What distinguishes these constitutive inferences from the rest?
- Which of the inferences that involve a lexical item are analytic?
- Assume that some lexical expressions have perceptual criteria of application (roughly equivalent: Assume that some lexical items express "recognitional" concepts). Which expressions are these? Under what conditions is a "way of telling" whether an expression applies constitutive of the identity of the expression?

These are all interesting and important questions, and you will be unsurprised to hear that I don't know how to answer them. I do, however, have a constraint to offer which, I'll argue, does a fair amount of work; in particular, it excludes many of the proposed answers that are currently popular in philosophy and cognitive science, thereby drastically narrowing the field. Here's the constraint: Nothing is constitutive of the content of a primitive linguistic expression except what it contributes to the content of the complex expressions that are its hosts; and nothing is constitutive of the content of a complex expression except what it inherits from (either its syntax or) the lexical expressions that are its parts.

Short form (*Principle P*): The constitutive properties of a linguistic expression qua linguistic include only its *compositional* properties.[2]

Principle P can't, of course, be a *sufficient* condition for content constitutivity. Suppose, for example, that all cows are gentle. Then, all brown cows are gentle a fortiori, so "cow" contributes *gentle* to its (nonmodal) hosts, and "brown cow" inherits *gentle* from its constituents. It doesn't follow—and, presumably, it isn't true—that *gentle* is constitutive of the content of either "brown," "cow," or "brown cow." The situation doesn't change appreciably if you include modal hosts, since not all necessary truths are analytic. That is, they're not all constitutive of the content of the expressions that enter into them. "Two" contributes *prime* to "two cows"; necessarily, two cows is a prime number of cows. But I suppose that *prime* isn't part of the lexical meaning of "two." Not, anyhow, if concept posession requires the mastery of whatever inferences are content-constitutive.

Still, P is a serious constraint on a theory of content, or so I maintain. For example: If P is true, then probabilistic generalizations can't be part of lexical content. Suppose that if something is a cow, then it is probably gentle. It doesn't follow, of course, that if something is a brown cow, then it is probably gentle, so "cow" doesn't contribute *probably gentle* to its hosts. Mutatis mutandis for other probabilistic generalizations; so probabilistic generalizations aren't part of lexical content.

Likewise, chapter 4 argued that if you grant principle P it follows, on weak empirical assumptions, that the *epistemic* properties of lexical expressions (e.g., the criteria for telling whether they apply) can't be among their essential properties. The argument went like this: Suppose that for modifier A, W is the way of telling whether A applies; and that AN is a complex expression containing the head N with A as its modifier; and that W^* is the way of telling whether AN applies. Then, by principle P, W must be part of W^*; the way of telling whether A applies must be part of the way of telling whether AN applies.[3] Arguably this works fine for words like, for example, "triangle" with respect to hosts like, for example, "red triangle," since if counting the sides is a good way of telling whether something is a triangle, it's likewise part of a good way of telling whether it's a red triangle. Nothing is *either* a triangle *or* a red triangle unless it's got three sides. But the suggestion that hosts inherit ways of telling from their constituents, though arguably it's okay for "red" in "red triangle," doesn't work, even prima facie, in the *general* case; for example, it doesn't work for "fish" in "pet fish."[4] Finding out whether it lives in a stream or lake or ocean is a good way of telling whether something is a fish; but it's a rotten way of telling whether it's a pet fish. Pet fish generally live in bowls. It follows, if principle P is true, that being a way of telling whether an expression applies is *not* an essential property of that expression. All forms of semantic empiricism to the contrary notwithstanding.

All this, I continued to argue in the earlier paper, should strike you as unsurprising. A way of telling whether an expression E applies to an object O is, at best, a technique (procedure, skill, or whatever) which allows one to tell whether E applies to O, *given that O is a good instance of E and the circumstances are favorable.* This must be right if there's to be any hope that ways of telling are possession conditions. I guess I know what "triangle" means. But it's certainly not the case that I can tell, for an *arbitrary* object in an *arbitrary* situation, whether "triangle" applies to it. (Consider triangles outside my light cone.) At best, my knowing what "triangle" means is (or requires) my knowing how to apply it to a good instance of triangles in circumstances that are favorable for triangle recognition. I don't think anybody disputes this. And, I don't think that anybody should.

But now, *the property of being a good instance doesn't itself compose.* What's a good instance of a fish needn't be a good instance of a pet fish, or vice versa. For that matter, what's a good instance of a triangle needn't be a good instance of a red triangle, or vice versa. That *goodinstancehood* doesn't compose is, I think, the ineliminable fly in the empiricist's ointment.

Notice, crucially, that *goodinstancehood*'s not composing does *not* mean that "pet fish" or "red triangle" aren't themselves compositional. To the contrary, O is a pet fish iff it's a pet and a fish, and O is a red triangle iff it's red and a triangle; "pet fish" and "red triangle" are thus as compositional as anything can get. What it means, rather, is that the epistemic properties of lexical items aren't essential to their identity qua linguistic. Not, anyhow, if principle P is true and only the compositional properties of an expression are among its constitutive properties.

The constitutive properties of an expression include only the ones it contributes to its hosts. But, in the general case, expressions don't contribute their good instances to their hosts (being a good instance of a fish isn't necessary for being a good instance of a pet fish). Since "criteria" (and the like) are ways of recognizing good instances, it follows that criteria (and the like) aren't constitutive properties of linguistic expressions. I think that's a pretty damned good argument that the epistemic properties of lexical items aren't constitutive of their identity qua linguistic. However, I have shown this argument to several philosophical friends who disagree. They think, rather, that it's a pretty damned good argument that principle P can't be true.[5] In particular, so the reply goes, epistemic properties, such as having some particular criteria of application, are essential to morphosyntactically *primitive* linguistic expressions (like "red" and "fish") but *not* to their hosts (like "red triangle" and "pet fish") *even in cases where the hosts are semantically compositional.*[6] If this reply is right, then a fortiori, the constitutive properties of a linguistic expression can't be among the ones that its hosts inherit.

I was, for reasons that the chapter 4 elaborated, surprised to hear this suggestion so widely endorsed. In particular, I argued like this: Consider *any* property P that is constitutive of E but not inherited from E by its hosts; I'll call such a property an "extra." If E has such an extra property, then, presumably, it would be possible for a speaker to satisfy the possession conditions for a complex expression containing E *without satisfying the possession conditions for E itself:* Somebody who has learned the linguistically essential properties of "pet fish," for example, need not have learned the linguistically essential properties of "pet" or "fish." For, by assumption, the mastery of "pet" and "fish" requires an ability to recognize good instances of each in favorable circumstances; whereas, again by assumption, the mastery of "pet fish" requires neither of these abilities (not even if it does require an ability to recognize good instances of pet fish in favorable circumstances).[7]

But, I supposed, it is something like true by definition that mastering a complex expression requires mastering its constituents, since, after all, constituents are by definition *parts* of their hosts. So, to say that you could master "pet fish" without mastering "pet" (or, mutatis mutandis, "red triangle" without mastering "red") is tantamount to saying that "pet" isn't *really* a constituent of "pet fish" after all; which is, in turn, tantamount to saying that "pet fish" is an idiom. Which, however, "pet fish" patently is not. So I win.

Now, there is a reply to this reply. ("Dialectics," this sort of thing is called.) One could just *stipulate* that recognitional capacities (or, indeed, any other sort of extra that you're fond of) are to count as constitutive of the primitive expressions that they attach to *even though* they are not inherited by the hosts of which such primitives are constituents. To which reply there is a reply once again; namely, that explanation is better than stipulation. Whereas principle P *explains* why meeting the posession conditions for a complex expression almost always[8] requires meeting the possession conditions for its constituents, the proposed stipulation just *stipulates* that it does; as does *any* account of constituency that allows primitive expressions to have extras.

To which there is again a reply: Namely, that you can't expect a theory to explain everything; and, given a forced choice between empiricism and the identification of the constitutive properties of an expression with its compositional properties, one should hold on to the empiricism and give up principle P. If that requires a revision of the notion of constituency, so be it. That can be stipulated too, along the lines: A (syntactic) part of a complex host expression is a *constituent* of the expression only if it contributes all of its semantic properties to the host *except* (possibly) *its epistemic ones.*

Thus far has the World Spirit progressed. I don't, myself, think well of philosophy by stipulation; but if you don't mind it, so be it. In this paper, I want to float another kind of defense for principle P; roughly, that the learnability of the lexicon demands it. I take it that these two lines of argument are mutually compatible; indeed, that they are mutually reinforcing.

Compositonality and Learnability

My argument will be that, given the usual assumptions, learnability requires that primitive expressions (lexical items) have no extras. A fortiori, it can't be that criteria, recognitional capacities, etc., are constitutive of primitive linguistic expressions but not inherited by their hosts.

The "usual assumptions" about learnability are these:

(i)   The languages with whose learnability we are concerned are semantically infinite (productive); that is, they contain infinitely many semantically distinct expressions.

(ii)   A theory of the learnability of a language is a theory of the learnability of the grammar of that language. Equivalently, a learnability theory has the form of a computible function (a "learning algorithm") which takes any adequate, finite sample of L onto a correct grammar G of L. I assume, for convenience, that G is unique.

(iii)   A learning algorithm for L is "adequate" only if the grammar of L that it delivers it *tenable.* A grammar G of L is tenable iff L contains not more than finitely many expressions that are counterexamples to G (and all of these finitely many counterexamples are idioms. See fn.8).

Comments:

• Assumption (i) is inessential. As usual, systematicity would do as well as productivity for any serious polemical purposes. (See Fodor and Pylyshyn, 1988.)

• Assumption (ii) is inessential. It's convenient for the exposition to assume that learning a language is learning a theory of the language, and that the relevant theory is a grammar of the language. But nothing turns on this. If, for example, you hold that language learning is "learning how" rather than "learning that," that's perfectly okay for present purposes; the kind of argument I'm going to run has an obvious reformulation that accomodates this view. I myself am committed to the idea that learning the semantics of the lexicon is *neither* "learning that" *nor* "learning how"; it's becoming causally (nomologically) connected, in a certain information-engendering way, to the things that the item applies to. That view of language learning is also perfectly okay with the present line of argument, I'm glad to report.

• Though I've borrowed the style of i–iii from learnability theory, it's important to see that the intuition they incorporate is really quite plausible independent of any particular theoretical commitment. Suppose that as a result of his linguistic experience, a child were to arrive at the following view (subdoxastic or explicit) of English: the only well-formed English sentence is "Burbank is damp." Surely something has gone wrong; but what exactly? Well, the theory the child has learned isn't tenable. English actually offers infinitely many counterexamples to the theory the child has arrived at. For not only is "Burbank is damp" a sentence of English, but so too are "Burbank is damp and dull," "The cat is on the mat," "That's a rock," and so on indefinitely. The core of i–iii is the idea that a learning algorithm that permits this sort of situation is ipso facto inadequate. A technique for learning L has, at a minimum, to deliver an adequate representation of the productive part of L; and to a correct representation of the productive part of a language there cannot be more than finitely many counterexamples among the expressions that the language contains.

• Assumptions (i–iii) constrain *language* learning rather than *concept* learning. I've set things up this way because, while I'm sure that lexicons are learned, I'm not at all sure that concepts are. What I'm about to offer is, in effect, a transcendental argument from the premise that a lexicon is learnable to the conclusion that none of the properties of its lexical items are extras. Now, historically, transcendental arguments from the possibility of language learning are themselves typically vitiated by empiricist assumptions: In effect, they take for granted that word learning involves concept acquisition (that, for example, the questions "how does one learn [the word] 'red'?" and "how does one learn [the concept] RED?" get much the same answer). This begs the question against all forms of conceptual nativism; and I'm inclined to think that some or other sort of conceptual nativism is probably true. (For discussion of the methodological situation, see Fodor and Lepore, 1992.)

"Transcendental arguments from language learning are typically vitiated by empiricist assumptions" might do well as the epitaph of analytical philosophy. I propose, therefore, to consider seriously only those transcendental conditions on language learning that persist *even if it's assumed that concept acquisition is prior to and independent of word learning.* That is, I will accept only such transcendental constraints as continue to hold even on the assumption that learning the lexicon is just connecting words with *previously available* concepts. I take it that constraints of that sort really *must* be enforced. Even people like me who think that RED is innate agree that *"red" expresses RED* has to be learned.

So, here's the argument at last.

Consider, to begin with, "pet fish"; and suppose that *good instances of fish are R* is part of the lexical entry for "fish": According to this lexicon, *being R* is part of the content that the word "fish" expresses (it's, if you prefer, part of the concept FISH). Finally, suppose that a specification of R is *not* among the contributions of "fish" to "pet fish" (*being R* is *not* part of the concept PET FISH), so R is an extra within the meaning of the act. Then, on the one hand, the lexicon says that it's constitutive of "fish" that *being R* is part of what tokenings of "fish" convey; but, on the other hand, the tokenings of "fish" in "pet fish" do not convey this. So "pet fish" is a counterexample to this lexicon.

Notice that the learnability of "fish" is not impugned by these assumptions so far; not even if principle P is true.[9] Tenability requires that English should offer *not more than finitely many exceptions* to the lexical representation of its primitive expressions. But, so far, all we've got is that English offers *one* exception to the lexical entry that says that "fish" conveys *being R*; namely, "pet fish," which typically does not convey *being R*. The most we're entitled to conclude is that, on the present assumptions, the child couldn't learn "fish" from data about "pet fish." Which maybe is true (though, also, maybe it's not).

However, an infinity of trouble is on the way. For it's not just the case that good instances of fish needn't be good instances of pets; it's also the case that good instances of big pet fish needn't be (for all I know, typically aren't) good instances of pets or of fish *or of pet fish.* Likewise, good instances of big pet fish that are owned by people who also own cats needn't be (for all I know, typically aren't) good examples of fish, or of pet fish, or of big pet fish, or of big pet fish that are owned by people who also own cats and who live in Chicago.

And so on, forever and forever, what with the linguistic form $(A^*N)_N$[10] being productive in English. The sum and substance is this: The good instances of the nth expression in the sequence A^*N are not, in general, inherited from the good instances of expression $n-1$ in that sequence; and they are not, in general, contributed to the good instances of expression $n+1$ in that sequence. Informally: you can be a good instance of N but not a good instance of AN; "pet fish" shows this. But, much worse, you can be a good instance of AN but not a good instance of $A(AN)_N$; and you can be a good instance of $A(AN)_N$ but not a good instance of AN. Good instances don't, as I'll sometimes say, *distribute* from modifiers to heads or from heads to modifiers, or from modifiers to one another. And their failure to do so is iterative; it generates an *infinity* of counterexamples to the compositionality of *goodinstancehood*.[11] Nothing in this depends on the specifics of the example, nor is the construction "A^*N" semantically eccentric. Adverbs, relatives, prepositional phrases, and indeed all the other productive forms of modification, exhibit the same behavior mutatis mutandis. Thus good instances of *triangle* needn't be good instances of *red triangle*; and good instances of *red triangle* needn't be good instances of *red triangle I saw last week*; and *good instances of red triangles I saw last week in Kansas* needn't be good instances of things I saw last week, or of triangles, or of red triangles, or of things I saw last week in Kansas, etc. The hard fact: *Goodinstancehood* doesn't distribute and is therefore not productive.

Every primitive expression has an infinity of hosts to which it fails to contribute its good instances (mutatis mutandis its sterotype, its prototype, etc.). So, a grammar that takes the ability to recognize *good instances* to be a possession condition for primitive items faces an infinity of counterexamples. As far as I know, this point has been missed just about entirely in the discussion of the "pet fish" problem in the cognitive science literature (to say nothing of the philosophical literature, where the problems that compositionality raises for epistemic theories of meaning have hardly been noticed at all). In consequence there's been a tendency to think of the semantics of "pet fish," "male nurse," and the like as idiosyncratic. It's even been suggested, from time to time, that such expressions are idioms after all. But, in fact, the pet fish phenomenon is completely general. That typical pet fish aren't typical pets or typical fish is a fact not different in kind from the fact that expensive pet fish are, practically by definition, not typical pet fish; or that very, very large red triangles aren't typical red triangles (or typical red things; or typical triangles). To repeat the point: it's not just that there are *occasional* exceptions to the compositionality of *goodinstancehood*. It's that there is an *infinity* of exceptions—which, of course, assumption (iii) does not permit.

Short form of the argument: The evidence from which a language is learned is, surely, the tokenings of its expressions. So, if a lexical expression is to be learnable, its tokenings must reliably manifest its constitutive properties;[12] that's just a way of saying what tenability demands of learnability. But tokenings of A(AN) do not, in general, reliably manifest the constitutive properties of AN on the assumption that the epistemic properties of A (or of N) are constitutive. Nothing about tokenings of "pet fish" reliably manifests the properties of good instances of fish (or of pets) tout court; nothing about triangles I saw in Kansas reliably manifests the properties of good instances of triangles tout court. And so on for infinitely many cases. So either tenability doesn't hold or the epistemic properties of lexical expressions aren't constitutive. Q.E.D.

Very short form of the argument: Assume that learning that fish typically live in lakes and streams is part of learning "fish." Since typical pet fish live in bowls, typical utterances of "pet fish" are counterexamples to the lexical entry for "fish." This argument iterates (e.g., from "pet fish" to "big pet fish" etc.). So a lexicon that makes learning where they typically live part of learning "fish" isn't tenable. Since the corresponding argument can be run for any choice of a "way of telling" (indeed, for any "extra" that you choose), the moral is that an empiricist lexicon is ipso facto not tenable. Q.E.D. again.

So, maybe we should give up on tenability? I really don't advise it. Can one, after all, even make sense of the idea that there are infinitely many sentences of English (infinitely many sentences generated by the correct grammar of English) which, nevertheless, are counterinstances to the mental representation of English that its speakers acquire? What, then, would their *being* sentences of English consist in? Deep metaphysical issues (about "Platonism" and such) live around here; I don't propose to broach them now. Suffice it that if you think the truth makers for claims about the grammars of natural languages are psychological (as, indeed, you should), then surely you can't but think that the grammars that good learning algorithms choose have to be tenable.

But I prefer to avoid a priorism whenever I can, so let's just assume that *being a tenable theory of L* and *being a well-formed expression of L* are not actually interdefined. On that assumption, I'm prepared to admit that there is a *coherent* story about learning English according to which recognitional capacities for fish are constitutive of the mastery of "fish" but not of the mastery of "pet fish" or of any other expression in the sequence A^* *fish*. That is, there's a coherent story about leanability that doesn't require tenability, contrary to assumption (iii) above.

Here's the story: The child learns "fish" (and the like) *only* from atomic sentences ("that's a fish") or from such complex host expressions as happen to inherit the criteria for fish. (Presumably these include such expressions as: "good instance of a fish," "perfectly standard example of a fish," and the like.) Occurrences of "fish" in any *other* complex expressions are ignored. So the right lexicon is *untenable* (there are infinitely many occurrences of "fish" in complex expressions to which "fish" does *not* contribute its epistemic properties). But it doesn't follow that the right lexicon is unlearnable. Rather, the lexical entry for "fish" is responsive to the use of "fish" in atomic sentences and to its use in those complex expressions which do inherit its epistemic properties (and only to those).[13]

But while I'm prepared to admit that this theory is *coherent*, I wouldn't have thought that there was any chance of its being *true*. I'm pretty sure that tenability is de facto a condition for a learning algorithm for a productive language. Here are my reasons.

I just said that, in principle, you could imagine *some* data about complex A^*N expressions constraining the lexical entry for N (and A) even if the lexicon isn't required to be tenable. The criteria of application for "perfectly standard example of a fish" are presumably inherited from the criteria of application for "fish." But, though this is so, it's no use to the child since he has no way to pick out the complex expression tokens that *are* relevant from the ones that aren't. The criteria for "perfectly standard fish" constrain the lexical entry for "fish"; the criteria for "pet fish" don't. But you have no way to know this *unless you already know a lot about English;* which, of course, the child can't be supposed to do.

What it really comes down to is that if the lexical representations that the learning procedure chooses aren't required to be tenable, then it must be that the procedure treats as irrelevant all data except those that derive from the behavior of *atomic* sentences. These are the *only* ones that can be un-question-beggingly relied on not to provide counterexamples to the correct entry. If *they typically live in streams or oceans* is specified in the lexical entry for "fish," that has to be learned from tokenings of sentences which reliably manifest that typical fish live in streams or oceans. And, if you don't know the semantics, so all you've got to go on is *form*, "a is a fish" is the only kind of sentence that can be relied on to do so. So it's the only kind of sentence whose behavior the child can allow to constrain his lexical entry for "fish."

You might put it like this: Compositionality works "up" from the semantic properties of constituents to the semantic properties of their hosts. But learnability works "down" from the semantic properties of hosts to those of their constituents. If the learning procedure for L is required to choose a *tenable* theory of L, then, with only a finite number of (idiomatic) exceptions, every host can (potentially) provide data for each of its lexical constituents. If, however, the learning procedure is *not* required to chose tenable grammars, then it has to filter out (ignore) the data about indefinitely many host constructions. The only way it can do this reliably is by a constraint on the *syntax* of the data sentences since, by assumption, their *semantics* isn't available; it's what the algorithm is trying to learn. And since, qua "extra," the property of being a good instance doesn't distribute from heads to modifers, the only syntactic constraint that an empiricist learning algorithm can rely on is to ignore everything but atomic sentences.

So then: If a representation of *goodinstancehood* is an extra, English *couldn't be learned* from a corpus of data that includes no syntactically atomic sentences. Notice that this is quite different from, and *much* stronger than, the claim that the child couldn't learn his language from a corpus that includes no *demonstrative* sentences. I would have thought, for example, that routine employments of the method of differences might enable the learning procedure to sort out the semantics of "red" given lots and lots of examples like "this is a red rock," "this is a red cow," "this is a red crayon." etc., even if a child had no examples of "this is red" tout court. But not so if mastering "red" involves having a recognitional capacity for red things. For even if, as a matter of fact, red rocks, red cows, and red crayons *are* good instances of red things, *the child has no way of knowing that this is so.* A fortiori, he cannot assume that typical utterances of "red rock," "red cow," etc. are germane to determining the epistemic clauses in the lexical entry for "red."

Here's the bottom line. Chapter 4 argued that an empiricist can't explain why everybody who understands "*AN*" also understands "*A*"; he can, at best, stipulate that this is so. The present argument is that, since epistemic properties turn on the notion of a good instance, and since being a good instance doesn't distribute from modifiers to heads or vice versa, if you insist on epistemic properties being specified in the lexicon you will have to treat them as extras. But the learnability of the lexicon is now in jeopardy; for the learning procedure for a lexical item has, somehow, to ignore indefinitely many expressions that contain it. In practice this means ignoring everything except atomic expressions, so the empiricist is committed to an empirical claim which is, in fact, very not plausible: You can only learn a language from a corpus that contains syntactically atomic sentences. There is, to repeat, no reason in the world to suppose that this consequence is true. Do you really want to be bound to a theory of lexical content which entails it? And, anyhow, what business has a metaphysical theory about meaning got legislating on this sort of issue?

However, I suppose most of the readers of this stuff will be philosophers, and philosophers like demonstrative arguments. And the argument I've given isn't demonstrative, since as it stands, it's compatible with a learnability theory that assumes that there are (adequately many) atomic sentences in the corpora from which language is learned. Well, I can't give you a demonstrative argument; but I can give you a diagnosis, one that I think should satisfy even a philosopher. To wit:

"Being able to tell" whether an expression applies is being able to recognize good instances. But being able to recognize a good instance of N doesn't mean being able to recognize a good instance of AN. That's because your ability to recognize a good instance of N may depend on good instances of N being R; and, since *goodinstancehood* doesn't compose, there's no guarantee that if good instances of N are R, then good instances of AN are also R. That is, as I remarked above, the empiricist's bane. Well, but just *why* doesn't *goodinstancehood* compose? Not, I think, because of some strange or wonderful fact about semantic properties that lexical items fail to share with their hosts, but rather because of a boring and obvious fact about typicality. What makes something a typical member of the set of Xs needn't be, and generally isn't, what makes something a typical member of some arbitrary sub- (or super-) set of the Xs. And even when it is, it's generally a contingent fact that it is; a fortiori, it isn't a necessary truth that it is; a fortiori, it isn't a linguistic truth that it is, since, I suppose, linguistic truths are necessary whatever else they are. Whether being red makes a triangle typical of the kind of triangles they have in Kansas depends on brute facts about what kinds of triangles they have in Kansas; if you want to know, you have to go and look. So it's perfectly possible to know what makes something a typical X, *and to know your language*, and nonetheless not have a clue what makes something a typical member of some sub- (or super-) set. *What makes a such and such a good example of a such and such just isn't a question of linguistics*; all those criteriologists and all those paradigm case arguments to the contrary notwithstanding.

It looks alright to have recognitional (or other epistemic) features that are constitutive of terms (or of concepts) until you notice that *being a good instance* isn't compositional, whereas the constitutive (a fortiori, the semantic) properties of linguistic expressions have to be. There is, in short, an *inherent* tension between the idea that there are recognitional terms or concepts and the idea that the semantics of language or thought is productive. That this sort of problem should arise for empiricists in semantics is perhaps not very surprising: Their motivations have always had a lot more to do with refuting skeptics or metaphysicians than with solving problems about linguistic or conceptual content. Empiricist semanticists took their empiricism *much* more seriously than they took their semantics; it was really justification that they cared about, not language or thought. Now all that playing fast and loose with the intentional has caught up with them. Good! It's about time!

I'm afraid this has been a long, hard slog. So I'll say what the morals are in case you decided to skip the arguments and start at the end.

The methodological moral is: *compositionality, learnability,* and *lexical entry* need to take in one another's wash. Considerations of learnability powerfully constrain how we answer the galaxy of questions around Q. And considerations of compositionality powerfully constrain the conditions for learnability.

The substantive moral is: All versions of empiricist semantics are false. There are no recognitional concepts; not even RED.

Notes

1. For all the cases that will concern us, I'll assume that concepts are what linguistic expressions express. So I'll move back and forth between talking about concepts and talking about words as ease of exposition suggests.

2. For our present purposes, Principle P can be taken to be equivalent to the "premise P" of chapter 4. The difference, in case anyone cares, is that whereas the current version constrains concept *constitution,* the other constrains concept *possession.* I suppose that whatever is required for having a concept should turn out to be constitutive of the concept, but perhaps not vice versa. If so, then some, but not all, of the constitutive properties of a concept will correspond to what chapter 4 called the "satisfiers for" the concept.

3. Of course, there will generally be *more than* one way of telling whether an expression applies; but there are obvious ways of modifying the point in the text to accomodate that.

4. It could be, of course, that this is because "red" and "triangle" express bona fide recognitional concepts but "pet" and "fish" do not. Notoriously, there isn't a lot of philosophical consensus about what words (or concepts) are good candidates for being recognitional. But, in fact, it doesn't matter. As will presently become clear, *however* the recognitional primitives are chosen, each one has *infinitely* many hosts to which it doesn't contribute its proprietary recognition procedure.

5. Philosophical friends who have endorsed this conclusion in one or another form, and with more or less enthusiasm, include: Ned Block, Paul Horwich, Brian Loar, Chris Peacocke, and Stephen Schiffer. That's a daunting contingent to be sure, and I am grateful to them for conversation and criticism. But they're wrong.

6. A variant of this proposal I've heard frquently is that, whereas the criterial and denotational properties of primitive expressions are *both* constitutive, only the denotational properties of primitive expressions are compositional; i.e., these are the only semantic properties that primitive expressions contribute to their hosts.

   It's more or less common ground that denotation (*instancehood,* as opposed to *good-instancehood*) does compose. The (actual and possible) red triangles are all found in the intersection of the (actual and possible) things "red" denotes and the (actual and possible) things "triangle" denotes; the pet fish are all found in the intersection of the things "pet" denotes with the things that "fish" denotes. And so on. It's thus in the cards that if, as I argue, compositionality rules out epistemic properties as linguistically constitutive, the default theory is that the constitutive semantic properties of words (and concepts) are their satisfaction conditions.

7. Notice that the argument now unfolding is neutral on the parenthetical caveat, i.e., it is *not* committed either way with respect to what Schiffer calls the "Agglomerative Principle." See chapter 4.

8. That is, always excepting idioms. This caveat isn't question-begging, by the way, since there are lots of independent tests for whether an expression is an idiom (e.g., resistance to syntactic transformation, to say nothing of not entailing and being entailed by its constituents).

9. On the other hand, if P *is* true, then this lexicon implies that "pet fish" is an idiom.

10. "$(A^*N)_N$" is a schema for the infinite sequence of (phrasal) nouns: $(A_1N)_N$, $(A_2N)_N$, etc.

11. Notice that the point I'm making does not depend on modal or "counterfactive" adjectives ("probably," "fake," etc.), where it's well known that AN can be compatible with, or entail, not-N.

    Notice, likewise, that though for convenience I talk of *goodinstancehood* being not compositional, exactly parallel considerations show that *constraints on* good instancehood aren't. The requirements for being a good instance of N aren't, in general, inherited by the requirements for being a good instance of AN; and so forth. This is a special case of the intertranslatability of stereotype-like theories of concepts (or lexical meanings) with exemplar-like theories.

12. It isn't good enough, of course, that the data from which an expression is learned should merely be *compatible* with its having the constitutive properties that it does. That "pet fish" is said of things that typically live in bowls *is* compatible with "fish" being said of things that typically don't; for it's contingent, and neither a priori nor linguistic, that people don't generally keep good instances of fish as pets.

13. Notice that the present proposal is not that the child learns that only primitive expressions exhibit the constitutive epistemic properties. "Typical triangle" and the like show that that's not true. The relevant constraint is (not on what the child learns but) on what data he learns it from.

Chapter 6

Do We Think in Mentalese? Remarks on Some Arguments of Peter Carruthers

For better or worse, I am much inclined to the view that the vehicle of more or less all[1] thought is an innate, nonnatural language. It has become the custom to call this putative language in which one thinks *Mentalese*, and I shall do so in what follows. I'm also much inclined to the view that the semantics of Mentalese is informational and atomistic—*informational* in the sense that Mentalese symbols mean what they do in virtue of their causal/nomic connections to things in the world; and *atomistic* in the sense that the metaphysically necessary conditions for any one Mentalese symbol to mean what it does are independent of the metaphysically necessary conditions for any other Mentalese symbol to mean what it does.

Actually, Carruthers (1996; all page references in this chapter are to this book) agrees with much of this. In particular, he thinks that the vehicle of thought is linguistic, and that a lot of concepts are probably innate. But he disagrees on two essential points: He holds that all *conscious* thinking is done in (as it might be) English;[2] so English is the language of conscious thought. And he denies that English has an atomistic semantics. Ergo, he denies that the language of conscious thought does.

Refreshingly, Carruthers doesn't claim that he has knock-down arguments for any of this; he holds only that his is the view of preference ceteris paribus. In particular, he takes it to be *introspectively plausible* that one thinks in English; and, if there are no decisive reasons for supposing that one doesn't, we ought to accept the introspectively plausible view. "All I need to show is that Fodor has no definitely convincing argument for the thesis that the language of all thinking is Mentalese, in order for the introspective thesis to win by default" (134). Accordingly, much of the first half of Carruthers's book is given to arguments that are supposed to show that the arguments that are supposed to show that we think in Mentalese aren't decisive.[3] These are the arguments I propose to discuss in what follows.

Not, however, that I disagree with Carruthers's conclusion. It's true that the arguments that we think in Mentalese aren't decisive. This is, quite generally, not the kind of area in which decisive arguments are available;

rather, it's the kind of area in which one has had a good day if one has managed not to drown. But I do think the arguments for the view that one thinks in Mentalese are pretty plausible; indeed, I think they are the most plausible of the currently available options. And I don't think that Carruthers is right about what he says is wrong with them.

Carruthers has three major points to make; one is about the relation between the Mentalese story and the Griceian program in semantics; one is about the relation between the Mentalese story and the notion of functional definition; and one is about the relation between the Mentalese story and semantic holism. I propose to consider each of these. First, however, a brief word or two about the introspection argument itself.

To begin with, it's not so introspectively obvious that it's introspectively obvious that one thinks in English. The psychologist Ken Forster once told me that when he eavesdrops on his thinking, what he hears is mostly himself saying encouraging things to himself. "Go on, Ken; you can do it. Pay attention. Try harder." My own thought processes, for what it's worth, are no more transparent, though the internal commentary that goes with them is generally less sanguine: "I can't solve this; it's too hard. I'm not smart enough to solve this. If Kant couldn't solve this, how can they possibly expect me to?" and so forth. It could be, of course, that people's introspections disagree.

Anyhow, as Carruthers is fully aware, it's surely not introspectively available that what one does when one talks to oneself is thinking; specifically, it's not introspectively available that what one finds oneself saying to oneself is what plays the causal/functional role of thought in the fixation of belief, the determination of action, and the like. Introspection itself *can't* show this, for essentially Humean reasons: Post hoc is one thing, propter hoc is another; causal connectedness can't be perceived, it has to be argued for. That is as true of the inner sense as it is of the outer.

Finally—though this is a long story, and not one I want to dwell on here—I sort of doubt that the theory that we think in English can actually be squared with the introspective data, even if we're prepared to take the latter at face value. The problem is this: "thinking in English" can't just be thinking in (sequences of) English words since, notoriously, thought needs to be ambiguity-free in ways that mere word sequences are not.

There are, for example, two thoughts that the expression "everybody loves somebody" could be used to think, and, so to speak, thought is required to choose between them; it's not allowed to be indifferent between the possible arrangements of the quantifier scopes. That's because, sans disambiguation, "everybody loves somebody" doesn't succeed in specifying something that is susceptible to semantic evaluation; and susceptibility to semantic evaluation is a property that thought has essentially. You can, to be sure, say in your head "everybody loves somebody" while remaining completely noncommittal as to which quantifier has which scope. That just shows that saying things in your head is one thing, and thinking things is quite another.

I take the moral to be that "everybody loves somebody" isn't, after all, a possible vehicle of thought. The closest you could come to thinking in English would be to think in some ambiguity-free regimentation of English (perhaps in formulas of what Chomsky calls "LF" (roughly, Logical Form). Maybe, for example, what's in your head when you think *that everybody loves somebody* on the interpretation where "everybody" has long scope, is "every$_x$ some$_y$ (x loves y)." That (give or take a bit) *is* the right sort of linguistic structure to be the vehicle of a thought. But (dilemma) it's surely not the sort of linguistic structure that is given to anybody's introspection; if it were, we wouldn't have needed Frege to teach us about bound variables. Maybe intuition does say that you think in word sequences; but there's good reason to think that it's wrong to say so.

It is, to put the same point a little differently, a defining property of mental representation theories to hold that the causal role of thoughts in mental processes is formally determined by the structure of their vehicles. This is where the representational theory of mind intersects the computational theory of mind, and I take it to be common ground between me and Carruthers. There are, however, two different causal roles that converge on the sequence of words "everybody loves somebody." That being so, this sequence of words corresponds to two different vehicles of thought *even assuming that we think in English*. What distinguishes the vehicles—what the difference in their causal/computational powers depends on—is their syntax.[4] And the syntax of English isn't introspectively available. Carruthers can have it that we think in English, or he can have it that we have introspective access to the vehicles of our thoughts; but he can't have it both ways.

Ambiguity isn't all that's wrong with English sentences as vehicles of thought, by the way; for sometimes they tell you not too little but too much. Was what you thought *that it was surprising that John arrived early*, or was it *that John's arriving early was suprising?* If you find that your introspections are indifferent, then either they are unreliable or you mustn't have been thinking in English since, of course, English distinguishes between these expressions. Carruthers holds that it's precisely the *conscious* thinking that you do in a natural language. So, if your thought that it was surprising that John arrived early *was* conscious, there *must* be an (introspectively available) fact of the matter about which sentence form you thought it in. Well, but is there?

However, let's put the introspection stuff to one side. I'm anxious to get to Carruthers's arguments against the arguments that we think in Mentalese.

First argument: Mentalese and the Grice program   Suppose that thoughts are relations to Mentalese sentences, and that the content of a thought is the meaning of the Mentalese sentence to which it is related. And suppose not only that Mentalese is a nonnatural language, but that it's an *ontologicaly independent* nonnatural language; that is, a Mentalese sentence's meaning what it does doesn't depend on any natural language sentence meaning what *it* does. What, on such a view, ought one to say about the meanings of natural language sentences themselves?

According to Carruthers, the thing for a friend of Mentalese to do would be to adopt "a Griceian approach to natural-language semantics, according to which sentence-meaning is to be explained in terms of the communication of beliefs.... [This] takes the notions of belief and intention for granted and uses them in explaining the idea of natural language meaning" (74).

However, as Carruthers rightly remarks, "Since Grice wrote, there has been a rich literature of counterexamples to [the Griceian] account" (77); and, probably, nobody now thinks that the reduction of the meaning of English sentences to facts about the communicative intentions of English speakers—or, for that matter, to any facts about mental states—is likely to go through. This is just a special case of the general truth that nothing ever reduces to anything, however hard philosophers may try. So, then, here's the way Carruthers sees the geography: The Mentalese story about thought is hostage to the Grice story about natural-language meaning; and the Grice story about natural-language meaning is a check that probably can't be cashed; so the Mentalese story about thought is, to that extent, in trouble.

However: Carruthers is wrong to think that it matters to Mentalese whether Grice can cash his checks. More generally, it doesn't matter to Mentalese whether *any* story about natural-language meaning can be made to work. More generally still, it doesn't matter to Mentalese whether there *is* such a thing as the meaning of a natural-language expression or (in case this is different) whether natural-language expressions have meanings. What Mentalese cares about—and all that it cares about that connects it with the Grice program—is that natural language is used to express thought.

What's required for natural language to express thought is that knowing a natural language can be identified with knowing, about each of its sentences, *which* thought it is used to express. It *might* turn out to be possible to parley this condition into a theory of natural-language meaning.

On the face of it, it's plausible enough that whatever expresses thought must ipso facto have a meaning that is determined by the content of the thought that it expresses. Cashing that intuition is, in effect, what Grice's theory of meaning was up to. But—to reiterate—neither the claim that we think in Mentalese, nor the claim that knowing a natural language is knowing how to pair its expressions with Mentalese expressions, requires that this be so.

Suppose, for example, that there is *no answer at all* to questions like "What does 'the cat is on the mat' mean?" just as there is no answer at all to questions like "What do tables and chairs mean?" or "What does the number 3 mean?" English sentences, numbers, and tables and chairs, are, by this supposition, not the sorts of things in which meaning inheres. Perhaps only thoughts or mental representations are. It wouldn't follow that there is no answer to questions like "What thought is 'the cat is on the mat' used to express?" In fact, it's used to express the thought that the cat is on the mat, whether or not it thereby *means* that the cat is on the mat; indeed, whether or not it means *anything*. To repeat: Neither thoughts having content, nor the use of language to express thoughts, depends on language having content. Not, at least, by any argument that I can think of; certainly not by any of the arguments that Carruthers has on offer.

But, if that's right, then the Mentalese story about thinking doesn't depend on the Griceian reduction going through after all. The most that could follow from the failure of Grice's reduction is that he was wrong to assume that whatever is used to express thought must mean something. It would, no doubt, be surprising and rather sad if Grice was wrong about this, since the intuition that sentences, words, and the like do mean something would have to be given up,[5] and it is a plausible intuition, as previously remarked.[6] But, once again, *the status of Mentalese wouldn't be in the least impugned.*

Here's a slightly different way to put this point. Grice had *a theory of linguistic communication* and *a theory of linguistic meaning*, and his idea was somehow to ground the latter theory in the former. Now, Grice's theory of linguistic communication is practically untendentious, at least for anybody who is a realist about intentional states. The theory was that you utter sentences intending, thereby, to indicate or express the content of the mental state that you're in; and your hearer understands you if he determines from your utterance the content of the mental state that you intended it to express or to indicate. I take this theory of communication to be practically platitudinously true. So far as I know, none of the myriad counterexamples to the Grice program that you find in the literature is an objection to his communication theory.

Notice, however, that this communication theory could be true even if mental states are the only things in the world in which content inheres. In particular, no commitments at all about natural-language meaning enter into it. This is not an accident; indeed, it's *essential* from Grice's point of view that his account of communication should *not* presuppose a notion of natural-language meaning. For his goal was to *derive* the theory of meaning from the theory of communication, and he had it in mind that the derivation should not be circular.

I suppose Mentalese does need Grice's communication theory; or, anyhow, something like it. An acceptable account of thought ought to say something about how thoughts are expressed; and if thoughts are what have content in the first instance, then it is natural to suppose that what communication communicates is the content of thoughts. But though it needs his theory of communication, Mentalese doesn't need Grice's theory of natural-language meaning; or, indeed, any theory of natural language whatsoever. For the Mentalese story is not just that the content of thought is prior to natural-language content *in order of explanation*; the Mentalese story is that the content of thought is *ontologically* prior to natural-language meaning. That is, you can tell the whole truth about what the content of a thought is *without saying anything whatever about natural-language meaning*, including whether there is any. Here, according to the Mentalese story, is the whole truth about the content of a thought: it's the content of the Mentalese sentence whose tokening constitutes the having of the thought. (This, by the way, is about as un-Wittgensteinian a view of the conceptual geography as it is possible to hold. If the Mentalese story about the content of thought is true, then there *couldn't* be a private language argument. Good. That explains why there isn't one.)

It may be just as well that Mentalese leaves all the questions about natural language semantics open, because—so it seems to me—it really might turn out that there is no such thing. Maybe all there is to what "cat" means is that it's the word that English speakers use to say what they are thinking about when they are thinking about cats. That, of course, isn't semantics in either of the familiar construals of the term: It doesn't specify either a relation between the word and its meaning or a relation between the word and what it's true of. It can't, however, be held simply as a point of dogma that English has a semantics; if you think that it does, then you will have to argue that it does. Grice failed to say what natural-language meaning is; and so too, come to think of it, has everybody else. Maybe natural-language is the wrong tree for a theory of meaning to bark up.

Bottom line: Carruthers thinks Mentalese is hostage to the Grice program. But it's not. So the failure of the Grice program isn't an argument against the claim that we think in Mentalese.

I turn, briefly, to Carruthers's last two arguments, both of which concern the doctrine that the semantics of Mentalese is atomistic.

To begin with, not every friend of Mentalese accepts this doctrine. Indeed, the standard view is that the meaning of Mentalese expressions is, at least partly, determined by their functional (inferential) roles. Carruthers himself holds that this is so, both for Mentalese and for English. But I have resisted assimilating the Mentalese story about thought to a functional role theory of meaning. That's because I believe that functional role semantics is intrinsically holistic and that a holistic semantics is incompatible with any serious intentional realism.

The argument for the second step is familiar: If semantics is holistic, then the content of each of your thoughts depends on the content of each of your others; and, since no two people (indeed no two time slices of the same person) have all their thoughts in common, semantic holism implies that there are no shared intentional states. Hence that there are no sound intentional generalizations. Hence that there are no intentional explanations. But intentional Realism *just is* the idea that our intentional mental states causally explain our behavior; so holistic semantics is essentially irrealistic about intentional mental states. So the story goes.

Carruthers has two points to make against this, neither of which seems to me to work, but both of which keep cropping up in the recent literature. Let's have a look at them.

First, though Carruthers admits that there are lots of differences between people's actual beliefs, he thinks that inferential role semantics, rightly formulated, can avoid concluding that there are no, or practically no, belief contents that people share. The idea is that "[the contents of] propositional attitudes are individuated in terms of their *potential* [sic] causal interactions with one another. These ... may remain the same across subjects whose actual propositional attitudes may differ.... [T]he same *conditionals* can be true of them" (111). So, for example, you know that today is Tuesday, so you infer that tomorrow is Wednesday. I don't know what day it is today, so I'm agnostic about what day it will be tomorrow. But that's okay because we do have this hypothetical in common: *if* we think today is Tuesday, *then* we think tomorrow is Wednesday. Since, according to the present proposal, the functional role of an attitude, and hence its content, is to be defined relative to these shared conditionals, the eccentricity of actual beliefs needn't imply that belief contents can't be shared. As I say, this sort of idea is getting around (cf. Block, 1993).

Well, it may help a little, but it surely doesn't help much. For one thing, belief hypotheticals shift along with the categoricals when indexicals are involved. It's because I believe that today is Tuesday that I believe that if I ask you, that's what day you'll say it is. If content identity is relativized to such hypotheticals, it's going to exhibit just the sensitivity to individual differences that Carruthers is trying to avoid.

And even where "eternal" propositions are at issue, it's a classic problem for functional role semantics that, on the one hand, what functional relations hold between your beliefs depends on what you think is evidence for what; and, on the other hand, what you think is evidence for what depends, pretty globally, on what else you happen to believe—that is, on what you *categorically* believe. You think Clinton is a crook because you've read that he is in the *Times*. I haven't read the *Times*, so we differ in categorical beliefs (viz., about what it says in the paper), so we can't share relevant Clinton-belief contents. Thus the old holism. Notice how little the move to hypotheticals does to help with this. What, for example, is the status, in the circumstances imagined, of the conditional: "If one has read the *Times*, then one thinks that Clinton is a crook." Answer: though it's true of you by assumption, it's quite likely to be false of me: If I'd read that about Clinton, dogmatic Democrat that I am, I would have concluded *not* that he's a crook but that the *Times* is unreliable. One man's modus ponens is another man's reductio, as epistemologists are forever pointing out.

Examples like "if you think it's Tuesday today, then you think it's Wednesday tomorrow," where belief hypotheticals are plausibly content-constitutive (and hence insensitive to, inter alia, the contents of contingent categoricals) are exceptional and therefore misleading. Rather, people who differ in their categorical beliefs are often likely to satisfy different belief hypotheticals *for that very reason*.[7] Bottom line: There is a short route from functionalism about meaning to meaning holism. As far as anybody knows, there's no way to get around that.

What's wrong with functional role semantics (FRS) is that it wants to analyze the content of a belief in terms of its inferential (causal) relations; whereas, plausibly, the direction of metaphysical dependence actually goes the other way 'round. The content of a belief determines its causal role; at least, it does in a mind that's performing properly. Holism is Nature's way of telling FRS that it has the direction of analysis backwards. None of this is discernably altered if FRS goes hypothetical.

Alright, alright, but doesn't that prove too much? "[Holism] is a problem for functional individuation generally, not a problem for functional role semantics in particular. So, unless we are prepared to accept that there really are no functionally individuated entities, we have as yet been given no reason for rejecting functional individuation of thoughts" (114). (See also Devitt, 1996, where this sort of line is pushed very hard.)

Well, but does avoiding the holistic embarrassments of FRS really require rejecting functional analysis per se? That, to be sure, would be a steep price; more than even people like me, who hate FRS root and branch, are likely to be willing to pay.

Notice, to begin with, that the claim that the individuation of *such and suches* is functional is, in and of itself, a piece of metaphysics pure and simple; in and of itself, it need have no *semantical* implications at all. Suppose, for example, I claim that hearts are functionally individuated; in particular, that hearts *just are* whatever it is that pumps the blood. Making this claim presumably commits me to a variety of metaphysical necessities; if it is true, then "hearts are pumps" and all of its entailments are among them. *But it commits me to nothing whatever that's semantic.* For example, it leaves wide open all questions about what, if anything, the word "heart" means; or about what the principles of individuation for the concept HEART are; or about what the possession conditions are for having that concept.

The reason that claiming that hearts are pumps leaves all this open is that you can't assume—not at least without argument—that if it's metaphysically necessary that *Fs* are *Gs*, then it follows that "*F*" means something about being *G*; or that knowing that *Fs* are *Gs* is a possession condition for the concept *F* (or for the concept *G*). Thus, for example: It's metaphysically necessary that two is the only even prime. It doesn't follow that "two" means *the only even prime* or that having the concept TWO requires having the concept PRIME, etc. Likewise, it's metaphysically necessary that water is H_2O. But it doesn't follow that "water" means H_2O; or that you can't have the concept WATER unless you have the concept HYDROGEN, etc.

Functionalism about hearts is, to repeat, a metaphysical thesis with, arguably, *no semantical implications whatever.* Whereas, *the holistic embarrassments for FRS arise precisely from its semantical commitments; in particular, from the constraints that it places on concept possession.* The intention of FRS is that if *I* is an inference that is constitutive of concept *C*, then having concept *C* requires in some sense acknowledging *I* (e.g., it requires finding *I* "primitively compelling," or it requires being disposed to draw *I* ceteris paribus, or whatever). Holism follows from this construal of concept possession together with the assumption (independently warranted, in my view) that there is no principled way to distinguish concept-constitutive inferences from the rest.

But, notice, this is *semantic* functionalism; it's functionlism about having the concept HEART (or about knowing what the word "heart" means). Semantic functionalism does lead to semantic holism, and by familiar routes. But ontological functionalism is fully compatible with semantic atomism. That's because, although functionalist theories of hearts bring all sorts of metaphysical necessities in their train, they place *no constraints whatever* on concept possession. *It's just a fallacy to infer the functional individuation of concepts from the functional individuation of the properties that they express.*

So it's alright to be a functionalist about hearts; at least it's alright for all the arguments that we've seen so far. But it's not alright to be a functionalist about conceptual content; that leads to holism, hence to unsatisfiable constraints on concept possession.

One last point on all of this: It is, for all I know, okay to be a functionalist about conceptual content if you can, at the same time, contrive to avoid being a semantic holist. And, of course, an FRS functionalist can avoid being a semantic holist if he is prepared to accept an analytic/synthetic distinction; which, in fact, Carruthers does. I don't want to argue about whether it's okay to philosophize on the assumption of an analytic/synthetic distinction; everyone has to sleep with his own conscience. Suffice it that you *don't* need an analytic/synthetic distinction—or any semantical views whatever—to be a functionalist about hearts and such; you just need to be an *essentialist*. Being an essentialist is quite a different thing from endorsing an analytic/synthetic distinction (unless there is reason to believe that truths about essences are ipso facto analytic—which there isn't because they aren't). That's another way to see why functionalism about concepts doesn't follow from functionalism about things; and why the former is so much the more tendentious doctrine. So much, then, for Carruthers's third argument.

I want to end by returning to a point I've already made in passing. I don't think that there are decisive arguments for the theory that all thought is in Mentalese. In fact, I don't think it's even true, in any detail, that all thought is in Mentalese. I wouldn't be in the least surprised, for example, if it turned out that some arithmetic thinking is carried out by executing previously memorized algorithms that are defined *over public language symbols for numbers* ("Now carry the '2,'" and so forth). It's quite likely that Mentalese co-opts bits of natural language in all sorts of ways; quite likely the story about how it does so will be very complicated indeed by the time that the psychologists get finished telling it.

But here's a bet that I'm prepared to make: For all our philosophical purposes (e.g., for purposes of understanding what thought content is, and what concept possession is, and so forth), nothing essential is lost if you assume that all thought is in Mentalese. Hilary Putnam once remarked that if you reject the analytic/synthetic distinction, you'll be right about everything except whether there is an analytic/synthetic distinction. Likewise, I imagine, in the present case: If you suppose that all thought is in Mentalese, you'll be right about everything except whether all thought is in Mentalese. More than that is maybe more than it's reasonable to ask for.

Appendix: Higher-Order Thoughts

I try never to think about consciousness. Or even to write about it. But I do think there's something peculiar about Carruthers's treatment. I quote (174): "Any mental state M, of mine, is conscious = M is disposed to cause an activated [i.e., not merely dispositional] conscious belief that I have M." Now, as Carruthers rightly notes, this formulation would appear to be circular on the face of it. So he immediately adds "so as to eliminate the occurrence of the word 'conscious' from the right hand side of the definition" the following:

Any mental state M, of mine, is conscious = M (level 1) is disposed to cause an activated belief that I have M (level 2), which in turn is disposed to cause the belief that I have such a belief (level 3) and so on; and every state in this series, of level n, is disposed to cause a higher-order belief of level $n + 1$.

But that can't really be what Carruthers wants. It eliminates the occurrence of the word "conscious" from the right hand side alright; but it does so precisely by leaving it open that each of the higher-order beliefs involved might be *unconscious*. And it is explictly Carruthers's view that having the *unconscious* n-level belief that one is having a certain $n - 1$ level thought is *not* sufficient to guarantee that the $n - 1$ level thought is conscious.

As far as I can see, two options are open to Carruthers, neither of which is satisfactory. He could simply reintroduce "conscious" to qualify each of the higher level beliefs (e.g., *M [level 1] is disposed to cause an activated conscious belief that I have M [level 2], etc.)*—thereby, however, failing to resolve the problem about circularity. Or he could deny that "conscious" really occurs in "conscious-level-n belief" for any level higher than 1; in effect, he could spell "conscious-level-n belief" with hyphens—thereby, however, raising the hard question why a merely conscious-with-a-hyphen level $n + 1$ belief should suffice to make a level-n belief conscious *without* a hyphen. Notice, in particular, that conscious-with-a-hyphen beliefs can't be assumed to be conscious sans phrase; if they were, the analysis would lapse, once again, into circularity.

Notes

1. This is to ignore a lot of subtleties; for example, whether we sometimes think in images, and the like. Such questions may matter in the long run, but not for the project at hand.
2. In what follows "English" is generally short for "English or some other natural language" and "sentence" is generally short for "sentence or some other expression." Nothing hangs on this except brevity of exposition.
3. Most of the second part of Carruthers's book is about consciousness. I say a little about that in the Appendix.

4. I gather from e-mails that Carruthers thinks what makes the difference between the two thoughts that everybody loves somebody is not the syntax of the representation that you token, but something like *the way you intend the token*. I'm not sure that I understand that. I don't think that I intend my thoughts at all; I think that I just have them. But, in any event, the suggested doctrine is pretty clearly not compatible with a computational theory of mind, so I don't want any.

5. I'm assuming that if "the cat is on the mat" is used to express the thought that the cat is on the mat, then if it means anything it means that the cat is on the mat.

6. Disquotation would still give *a* sense to the thought that sentences, words, and the like mean something; to that extent they would be better off than numbers and tables and chairs, for which conventions of (dis)quotation have not been defined. Cold comfort, if you ask me.

7. There are, of course, all sorts of other ways in which belief hypotheticals shift with belief categoricals. I am cynical; I think all politicians are crooks. So (given, as usual, reasonable rationality) it's true of me that if I think Francis is a politician, then I think Francis is a crook. You are naive; you think no politicians are crooks. So, it's not true of you that if you think Francis is a politician then you think Francis is a crook. If the contents of our beliefs are relativized to these hypotheticals, we get the usual dreary result: Our beliefs about politicians aren't contradictories.

Chapter 7

Review of A. W. Moore's *Points of View*

Proust's Swann is obsessed by what he doesn't know about Odette. His anguish has no remedy; finding out more only adds to what he does know about her, which isn't what's worrying him. Since Kant, lots of philosophers have suffered from a generalized and aggravated form of the same complaint. They want to know what the world is like when they aren't thinking about it; what things are like, not from one or other point of view, but "in themselves." Or, anyhow, they think that maybe that's what science aims to know, and they wonder whether it's a project that makes any sense. They are thus concerned with "the possibility of objectivity."

Moore's book is a sort of defense of the possibility of objectivity (what he often calls the possibility of "absolute representations"). He doesn't argue that objectivity is ever attained or ever likely to be; perhaps not even in our most scrupulous investigations. All he wants is that the goal should be coherent. Given the modesty of the enterprise, it's really shocking the conclusions that he's driven to. "We are shown that absolute representations are impossible.... [But] ... we do well to remind ourselves ... that what we are shown is nonsense. Properly to replace '*x*' in the schema "*A* is shown that *x*" is a quasi-artistic exercise in which one creates something out of the resonances of (mere) verbiage. There is no reason whatever why this should not sometimes involve making play with inconsistency ..." (272)

It's, anyhow, a fascinating story how Moore gets to this. I think, in fact, that it's a cautionary tale. Moore takes on board, from the beginning and essentially without argument, the currently received view of meaning in philosophy. The rest follows as the night the day. What's splendid about the book is that, unlike most philosophers who share his premises, Moore is prepared to face the consequents. What's appalling is the tenacity with which Moore clings to an account of representation from which it follows that his own philosophical views, inter alia, are nonsense. ("Inter alia" because, of course, your views and mine are nonsense too.) Not just *like* nonsense, mind you, but the real thing: on all fours with "Phlump jing ax," since "there can be no other reason for an utterance's failing to be a representation than that certain [of the] words lack meaning" (201). I propose to trace, briefly, the course of these events. At the end, I'll moralize a little about how things stand if Moore's book is read, as I think it should be, as a reductio of his theory of meaning.

The book starts oddly: "[A] representation need not be objective.... [A] representation can be from a point of involvement.... I shall call any [such] representation ... 'subjective'" (8). This makes representations that contain indexicals (like "It is humid today") perspectival and subjective, since "[I]f I wish to endorse an assertion I made yesterday of [this] sentence, I have no alternative but to produce a representation of some other type ('It was humid yesterday')" (12). This, to repeat, is an odd way to start. For although indexical sentences are uttered from a point of view (in the sense that "it's raining" can be true here and false there at one and the same time; or true then and false now in one and the same place), there doesn't seem to be anything particularly subjective about them. If what you want is subjectivity, try "red is warmer than green" or "Callas was better than Tebaldi."

Indexicality is about content, whereas objectivity is about truth. If you take indexicality as a paradigm of the perspectival, then it's a mistake to run "perspectival" together with "subjective." What's interesting about a sentence like "it's raining" is that the content of the assertion that I make when I say it depends on where and when I am. What's interesting about "red is warmer than green" is something quite different; namely, that (maybe) it can be true for me but false for you, or (maybe) there's no matter of fact at all about whether it's true. There's, anyhow, certainly a matter of fact about whether it's raining; so what on earth does whether a representation is indexical have to do with whether it's objective? So one wonders as one reads along.

I guess what Moore has in mind is something like this: The content of an indexical representation depends on its context. But so likewise do the contents of representations of every other kind since having a representation means having a language, and "having a language involves having an outlook; or, more specifically ... having a language involves having its own distinctive outlook" (94). Moore swallows whole not just Wittgenstein, but also Whorf and Saussure; which is, perhaps, a little quaint of him. So the content of a representation, indexical or otherwise, is fixed partly by its conceptual role, by the inferences it licenses. That one can only represent things from the point of view of one's inferential commitments is therefore part and parcel of what content is. But representations from a point of view are ipso facto perspectival, hence ipso facto not objective. It seems, in short, that semantics puts pressure on epistemology across the board; there is a prima facie conflict between the possibility of objectivity and the idea that content is holistic. How to resolve this conflict is what Moore's book is *really* about, though I doubt he'd want to put the case this way.

You don't have to resolve it, of course; you could try to just live with it. Quite a lot of philosophers, including some of our most eminent contemporaries, seem to think that's indeed the best that you can do. So Thomas Kuhn famously maintained that, because concepts are implicitly defined by the theories that endorse them, radically different theories are ipso facto "incommensurable." There is, to that extent, no such thing as a rational choice between conflicting theoretical paradigms, and there is no fact of the matter which, if either, of conflicting paradigms is objectively correct. Changing your paradigm isn't changing your mind, it's changing your lifestyle.

Variants of this disagreeable idea are on offer from philosophers with as little else in common as, say, Wittgenstein, Quine, Goodman, Putnam, Derrida, and Davidson, among many others. Part of what's wrong with it is that it's so unstable. Just how major does the disagreement in background commitments have to be before incommensurability sets in? And if, as seems likely, there is no principled answer to this question, isn't it the moral that *all* judgment turns out to be more or less subjective? We're not far from the view that argument and persuasion are mostly power politics in disguise; class politics on some accounts, sexual politics on others. Whatever became of the disinterested pursuit of objective truth?

There is, however, a traditional, more or less Kantian way to split the difference between what's perspective and what isn't; it's the doctrine called "transcendental idealism," and much of what's most interesting in Moore's book is about it. It's one thing, Moore says, to suppose that representation is ipso facto "from an outlook"; it's another to conclude that representation is ipso facto not objective. The second follows from the first only on the assumption that apparently conflicting perspectival representations aren't capable of being integrated from a point of view that embraces both.

Suppose one takes for granted that representation is, as Moore puts it, always "of what's there anyway." That is, all of our points of view are perspectives on the same world, the world to which our beliefs conform insofar as they are true. True beliefs have to be mutually consistent, so it must be possible, in principle at least, to integrate all of the true perspectives, even if only by conjoining them. At the end of the day, representation is still perspectival, but perhaps the only perspective to which it is *intrinsically* relative is our point of view as rational inquirers, or cognitive systems, or (as Moore sometimes puts it) as processors of knowledge. Maybe you can't know what things are like in themselves, but you can, at least in principle, know what they are like insofar as they are possible objects of knowledge. If this sounds a bit truistic, so much the better.

That's encouraging as far as it goes, but it doesn't really go far. For one thing (a point Moore doesn't emphasize), the perspectival relativists he is trying to split the difference with are unlikely to grant him his "basic assumption" of a ready-made world. To the contrary, it's part and parcel of the Kuhnian sort of view that thinkers with different paradigms ipso facto live in different worlds. That's the metaphysical obverse of the semantic claim that you can't integrate divergent paradigms. (Since Kuhn takes content to be paradigm-relative, what you get if you conjoin representations from different paradigms is just equivocations.) Indeed, the many-worlds story is, arguably, just the incommensurable-paradigms story translated from semantics to metaphysics; so, arguably, you can't have one without the other. If it's not clear how literally we're to take either, or even what taking them literally would amount to, that just shows what a mess we're in.

And second, a point to which Moore is entirely alert, transcendental idealism looks to be self-refuting. If you *really* can't say anything about the world except as it is represented, then one of the things that you can't say is that you can't say anything about the world except as it is represented. For the intended contrast is between how the world is *as represented* and how it is *sans phrase*. But how the world is sans phrase is how it is *not* from a perspective, and it's part of the story we're trying to tell that representation is perspectival *intrinsically*. It's worth remarking how much worse off Moore is than Kant in this respect. Kant thought that there are substantive constraints on the "form" of the possible objects of knowledge, and hence that we can't know about what fails to conform to these constraints. Moore, following the early Wittgenstein, thinks that there are substantive constraints on the form of the possible objects of representation, and hence that we can't even *think* about what fails to conform to these constraints. For Kant, transcendental idealism was strictly mere metaphysics (that is to say, its truth is unknowable); for Moore it is strictly mere babble (that is to say, its content is unspeakable). It is, I think, a general rule that whenever philosophy takes the "linguistic turn" the mess it's in gets worse.

Transcendental idealism is nonsense by its own standards. So be it: "[T]his leaves scope for all sorts of distinctions.... We can think of the Nonsense Constraint as offering the following guideline when it comes to making sense of the schema 'A is shown that x': namely to prescind altogether from considerations of content and to think more in aesthetic terms.... To say of some piece of nonsense that it is the result of attempting to express the inexpressible is *something like* [sic] making an aesthetic evaluation. It concerns what might be called, justly, if a little grandiloquently, the music of words" (202). One is powerfully reminded of Frank Ramsey's salubrious wisecrack about the embarrassing bits of the *Tractatus*: "What can't be said can't be said and it can't be whistled either."

I'll spare you the details, which are complicated and, in my view, much less than convincing. Roughly, we have lots of "inexpressible" knowledge. Its being inexpressible is somehow connected with its being knowing how (e.g., knowing how to process knowledge) rather than knowing that, though I find this connection obscure. Anyhow, sometimes we're driven to try to express the inexpressible (that we are has something to do with our aspiring to be infinite); when we do so, we talk nonsense. For example, we say things such as that representations can (or can't) be absolute. Since this is nonsense, saying it can, of course, communicate nothing. But that we are inclined to say it *shows* all sorts of things. For example, the nonsense we're tempted to talk about value shows that "things are not of value tout court. Nothing is. However, another thing we are shown is that they are of value tout court. Our aspiration to be infinite, precisely in determining that these things are of value for us, leads to our being shown this" (xiii). It's tempting to dismiss this sort of thing as merely irritating. But if you propose to do so, you have to figure out how to avoid it while nevertheless relativizing meaning to perspective and retaining a respectable notion of objectivity. It is enormously to Moore's credit that he has faced this squarely.

But, after all, where does having faced it get him? According to enemies of objectivity like Kuhn, it's false to say that the world chooses between our theories. According to friends of objectivity like Moore, it's nonsense to say that the world chooses between our theories. Objectivity might reasonably complain that with such friends it doesn't need existentialists. What's remarkable is that neither side of the argument considers, even for a moment, abandoning the premise that's causing the troubles; namely, that content is ipso facto perspectival and holistic.

Perspectival theories of meaning make objectivity look terribly hard. So much the worse for perspectival theories of meaning; objectivity is *easy*. Here's some: My cat has long whiskers. He doesn't (as Moore would be the first to agree) have long whiskers "from a perspective" or "relative to a conceptual system" or "imminently"; he has long whiskers *tout court*. That's what my cat is like *even when I am not there* (unless, however, someone should snip his whiskers while I'm not looking; which is not the sort of thing that Moore is worrying about). Prima facie, the ontological apparatus that's required for my cat to have long whiskers is extremely exiguous; all that's needed is my cat, long whiskers, and the state of affairs that consists of the former having the latter.

Likewise, it is possible for me to represent my cat's having long whiskers. Indeed, I just did. The semantic apparatus required for this is also exiguous, and for much the same reasons. All I need is to be able to represent my cat, long whiskers, and the state of affairs that consists of the former having the latter. First blush at least, neither my cat's having long whiskers nor my representing it as having them would seem to imply a perspective or an outlook or a desire for infinity, or anything of the sort; the objectivity of the fact is about all that the objectivity of the representation requires. So whence, exactly, all the angst?

That all this is so, and objectively so, is far more obvious than any general principles about the relativity of content to perspectives or conceptual systems. If, therefore, your perspectivist semantics leads you to doubt that it is objectively so, or that it can coherently be said to be objectively so, it's the semantics and not the cat that you should consider getting rid of. That this methodological truism should have escaped so many very sophisticated philosophers, Moore among them, seems to me to be among the wonders of the age.

Part III
Cognitive Architecture

Chapter 8

Review of Paul Churchland's *The Engine of Reason, The Seat of the Soul*

[W]e are now in a position to explain how our vivid sensory experience arises in the sensory cortex of our brains ... [and is] embodied in a vast chorus of neural activity ... to explain how the [nervous system and the musculature] perform the cheetah's dash, the falcon's strike, or the ballerina's dying swan.... [W]e can now understand how the infant brain slowly develops a framework of concepts ... and how the matured brain deploys that framework almost instantaneously: to recognize similarities, to grasp analogies, and to anticipate both the immediate and the distant future.
—Paul Churchland, *The Engine of Reason, The Seat of the Soul*, pp. 4–7

I do think that is naughty of Professor Churchland. For one thing, none of it is true. We don't know how the brain deploys its concepts to achieve perception and thought; or how it develops them; or even what concepts are. We don't know how the motor system contrives the integration of the lips, tongue, lungs, velum, and vocal cords in the routine utterance of speech, to say nothing of special effects like the approximation of dying swans by whole ballerinas. Nor do we know, even to a first glimmer, how a brain (or anything else that is physical) could manage to be a locus of conscious experience. This last is, surely, among the ultimate metaphysical mysteries; don't bet on anybody ever solving it.

Nor does Churchland's hyperoptimistic account of the current situation in "cognitive neuroscience" represent anything like the consensus view among practicioners. I'm sure this is so, because I've lately made a little game of reading the above and related quotations to friends in linguistics, cognitive psychology, physiological psychology, philosophy, and computer theory. I had hoped to pass along some of their comments for your amusement, but those who weren't dumbstruck used words that even *The New Yorker* still won't print.

What makes this naughty of Churchland, and not just ill advised, is that his book purports to be a popular introduction. Churchland wants to let the layman in on what's been going on in mind/brain studies, and on implications and applications all the way from science and technology to morality and jurisprudence. For example, not only do "[a]rtificial neural networks have the power to change the way we do science" (314) but "[the] new scanning techniques will allow us to accumulate a large database concerning individual profiles of brain function, from normals to violent sociopaths.... A large number of such pairs will constitute a training set for the sort of diagnostic network we desire, a network that, once trained, will accurately diagnose certain types of brain dysfunction and accurately predict problematic social behavior" (311).

If you're in the research business, you will recognize at once the rhetoric of technohype. It is the idiom of grant proposals and of interviews in the Tuesday *New York Science Times*: *The breakthrough is at hand; this time we've got it right; theory and practice will be forever altered; we have really made fantastic progress, and there is now general agreement on the basics; further funding is required.* Professionals are inured to this sort of stuff and write it, and write it off, as a matter of course. I hope that the laity will prove to be equally cynical. For Churchland's book is, in fact, not an overview but a potboiler. It's a polemic, generally informed and readable, for a certain quite parochial account of the methodology and substance of psychology: what they call in the trade "West coast" cognitive science. (Perhaps it goes without saying that "West coast," in this usage, denotes a spiritual rather than a geographical condition.)

Methodologically, Churchland endorses a research strategy for the study of cognition that is rooted in neurological modeling. This will seem to many like simply common sense. It's the brain that makes intelligence and consciousness; so if you want to know what intelligence and consciousness are, it's brains that you need to look at. Well, maybe so, but also maybe not. Compare: "If it's flight that you want to understand, what you need to look at is feathers." Actually, it's in the nature of research strategies that they get vindicated after the fact if at all. One way to study the mind/brain is to try to understand its chemistry and physiology. Another way is to try to simulate its characteristic behavioral effects. Another way is to forget about its behavioral effects and try to understand its cognitive competence. Another way is to construct experimental environments in which the details of its information processing can be selectively scrutinized. Another way is to examine its evolutionary precursors; or the character of its ecological adaptivity; or the course of its ontogeny, and so on; and there are surely lots of ways to study the mind/brain that no one's thought of yet. In fact, none of the strategies that we've tried so far has been extraordinarily successful, the ones that Churchland likes included. Probably the right thing to do is just what we are doing: work on all of them together and hope that something will turn up.

Methodology aside, the actual theory of the mind/brain that Churchland is pushing is a version of connectionism according to which perception and thought are carried out by "parallel distributed processors" that are organized into "neural networks." And his message is that connectionism is inevitable, so you had better learn to like it: "You came to this book assuming that the basic units of human cognition are states such as thoughts, beliefs, perceptions, desires, and preferences.... [By contrast] ... it is now modestly plain that the basic unit of computation is the *activation vector*. It is now fairly clear that the basic unit of computation is the *vector-to-vector* transformation. And it is now evident that the basic unit of memory is the *synaptic weight configuration* [p. 322].... It is now beyond serious doubt that this is the principal form of computational activity in all biological brains" (266). So, like, get with it.

Let us, nonetheless, do our best to keep our calm. To begin with, "neural network" is a name (not, in the first instance, for a network of neurons, but) for a certain kind of computer, the organization of which may, or may not, throw some light on some aspects of human cognition. It's a long story how these devices work; suffice it that they differ, in various ways, from the "classical" architecture of Turing machines, or von Neumann machines, or desktop computers. One of the differences is that networks don't exhibit the distinction between program and memory that is characteristic of the more familiar computational devices. Rather, both the current computational proclivities of a network and the residual effects of its computational history are encoded by connections between a very large number of relatively simple, switch-like elements.

Networks compute in parallel, by sending waves of activation through the connected elements. How much of the activation goes where depends on which connections are operative, and which connections are operative depends on the network's previous activations. You can make this picture look like the cutting edge of science if you think of the elements as neurons interactivating one another at synapses of variable resistance. But you can equally make it look quite dreary and recidivist: Think of the elements as "ideas" and call the connections between them "associations," and you've got a psychology that seems no great advance on David Hume. I'm sorry if you find that disappointing, but there is worse to come.

Since their computational histories affect their computational proclivities, networks are able to learn. Much has been made of this, but in fact it's a tautology, not a breakthrough. *Any* device whose computational proclivities are labile to its computational history is *thereby* able to learn, and this truism includes networks and classical machines indiscriminately. What distinguishes between them is that, although both can learn, the former can't be programmed but have to be trained. As it turns out, that's a mixed blessing.

Because computation in networks is massively parallel (in effect, the elements are interactivated simultaneously) they can sometimes solve problems very fast; that's the good news. But which computations a network is disposed to perform depends on the character of the connections between many, many elements. To set the strength of each of these connections "by hand" would be astronomically time consuming even if you know which settings would produce the computations that you want— which typically you don't and which there is no general procedure for finding out. That's the bad news. The discovery that parallel machines are likely to be fast but hard to program (hence, by the way, hard to debug if something goes wrong) is also not a breakthrough; it's a commonplace of the computer scientist's art. "West coast" cognitive science is betting that the benefits of massively parallel computing are probably worth the costs; "East coast" cognitive science is betting that they're probably not. In any event, networks are just one of the options; there are as many kinds of computational architectures as you like that are intermediates between purely serial and purely parallel machines. No doubt one of them (quite possibly not one that's now even imaginable) will turn out to be right for modeling the computations that the brain performs. That's all that anybody's got so far.

If that's all that anybody's got so far, then what is the fuss about? I mean, how do you promote *that* sort of situation into a scientific breakthrough of the first magnitude? Here's how. You hew to the main maxim of technohype: *What doesn't work, you don't mention.* Some illustrations from Churchland follow.

Classical machines "learn" inductive generalizations by executing software programs that extract statistical regularities from their input data. Networks "learn" inductive generalizations because their training alters their internal connections in ways that model such statistical regularities. (It typically requires a lot—I mean, *a lot*—of training to produce such alterations. That's the price that networks pay for massively parallel computation.) In either case, you end up with a representation of the statistical regularities in the input; so far, there is *no evidence at all* that the two kinds of machines differ in what such generalizations they can learn. Churchland is deeply impressed by how much better than people trained networks are at certain kinds of perceptual discrimination tasks (at distinguishing sonar pictures of mines from sonar pictures of rocks, for example). But that's the wrong comparison; all it shows is that people aren't very good at doing statistics in their heads. The right comparison is how networks fare in competition with conventional computers that are programmed with statistical analysis software. When you make such comparisons, it turns out (in any of the several cases I've heard of) that the two kinds of machines succeed in much the same tasks, but the networks typically are markedly slower. Churchland reports none of these results.

The worries that Churchland doesn't talk about go still deeper. If you want to ask whether the correlation of property *A* with property *B* is statistically reliable in a body of data, you have to *precode* the data for *A*-ness and *B*-ness. You can't assess the inductive likelihood that the next red thing that you come across will be a square unless you've got your previous experience coded for redness and for squareness. Often enough, this question of what to code for in the data is the hardest part of solving an inductive problem: The statistics of finding correlations is relatively straightforward once you know which properties to look for correlations between. Networks have this problem too, of course, and Churchland has no more idea of how to solve it than I do; that is, no idea at all. What he does instead is rename it, so what used to get called "properties" Churchland calls "dimensions." This too is not a breakthrough.

Some of the things that minds respond to do seem to have a natural taxonomy. Think of colors (one of Churchland's parade examples): You can construct a space with two dimensions of hue (red to green and blue to yellow) and one dimension of saturation (black to white), and each color will occupy a unique position in this space. That is, you can identify each color with the unique triple of numbers (a "vector") that gives its position on each of the dimensions. With such a scheme for coding the data in hand, it's duck soup to answer statistical questions like whether, in a certain sample of colors, the saturation is correlated with the hue.

Churchland is in love with dimensions, vectors, and prototypes. His idea is to do for concepts at large what hue and saturation allow you to do for colors: locate them in a "multidimensional vector space" and then look for correlations between the vector values. So, for example, your concept of a certain dog is maybe a vector of (proto)typical values along whatever the dimensions are that define dog-space; and your concept of dogness per se is maybe a vector of prototypical values along whatever the dimensions are that define animal-space. And so on. (Strictly speaking, vectors determine *points* in a vector space. But if, like Churchland, you think that concepts are too vague and fuzzy to be points, you can take them to be the regions around the points. Fuzzy is one of the things that West coast cognitive science likes best.)

So concepts are regions in a space of cognitive dimensions; *that's* the breakthrough. "What cognitive dimensions?" Well may you ask. You can get experimental subjects to estimate how similar and how different color samples are; and, if you analyze their answers, you can recover the perceptual dimensions of color. But colors are *special* in that respect; they really are all perceived as occupying positions in a space of a (relatively) few dimensions, which is to say that they are seen as similar and different in a (relatively) small number of respects. Surely, nothing of the sort is true *in general* of the things that our concepts apply to? Surely there isn't a dimensional space in which, as it might be, the color red, lawyers, Easter Sunday, and the number three are variously propinquitous? Red is more similar to orange than it is to green. Is it also more similar to Easter Sunday than it is to lawyers? If this question strikes you as a nonsense, beware the guy who tells you that concepts are vectors in cognitive space.

To be sure, it's always possible to contrive a space ad hoc; perhaps that's why Churchland seems so confident that things will turn out well for vector coding. For example, you could have a space one dimension of which represents *degrees of being Easter Sunday*. Easter Sunday could then be represented, by a vector in this space, as having a higher value on that dimension than anything else. If you've got only finitely many concepts to taxonomize, you will need only finitely many spatial dimensions to specify such vectors. (And if you've got infinitely many concepts, spaces of infinitely many dimensions are likewise available.) But, of course, this is just an unedifying kind of fooling around; it's an imperspicuous way of saying "what makes Easter Sunday different from lawyers is that Easter Sunday is Easter Sunday and lawyers aren't." That is not news, and the appeal to vectors, dimensions, and prototypes isn't doing any work.

Churchland writes as though he thinks that prototype vectors approximate a general answer to the abysmally deep question of how a mind represents the things that it thinks about. But the way he uses them, they don't; they're just a way of talking. So, having described the progress of mechanics from Aristotle to Einstein as the transformation of their respective prototype vectors, Churchland remarks that such achievements are continuous with "the very same activities of vector processing, recurrent manipulation, prototype activation, and prototype evaluation as can be found in some of the simplest ... cognitive activities ... [these being] the mechanisms embodied in a large and highly trained recurrent neural network" (121). The difference is just that Einstein's prototype "geodesic paths in a non-Euclidean space-time—is admittedly arcane to most of us" (120). That's how you say in technohypespeak that Einstein was smart enough to figure out that planetary orbits are straight lines in space-time but you and I probably wouldn't have been. As far as I can tell, that's all it says.

Saying that we conceptualize things as regions in a multidimensional space is saying no more than that we conceptualize things. The appearance of progress is merely rhetorical. If you want vectors and dimensions to be a breakthrough and not just loose talk, you've got to provide some serious proposals about *what the dimensions of cognitive space are*. This is exactly equivalent to asking for a revealing taxonomy of the kinds of concepts that people can have. Short of the puzzles about consciousness, there isn't any problem about the mind that's harder or less well understood. It goes without saying (or anyhow it should go without saying) that Churchland doesn't know how to provide such a taxonomy. Nor do I. Nor does anybody else.

Oh well, we've all heard psychology technohype before. Last time it was expert systems. The time before last it was operant conditioning and reinforcement theory. The time before that it was classical conditioning. The time before that it was the association of ideas. If you want a breath of reality, just to clear your head, look up "parallel distributed processing" in the fourth edition of Henry Gleitman's majesterial text *Psychology*. PDPs get about one page (out of eight hundred). Gleitman concludes: "The PDP approach is very ambitious ... [but] as yet it is much too early to come to any final evaluation.... [W]hatever we might say today will almost surely be out of date tomorrow" (291–292). Quite so. That's the way it goes with technohype.

The fact is (have I mentioned this before?) that we don't know how minds work, though we're pretty good at explaining certain aspects of perception. Nor do we know how brains work, if that's a different question, though we're getting pretty good at finding out *where* it is that they do their mental stuff. Churchland duly retails the latest gossip about MRIs and PETs and other such capital intensive hardware. This new technology may turn out to be useful for psychology, or again it may not. If you don't know *what* the carburetor is, you may not care all that much *where* the carburetor is. Ditto the neural locations of consciousness and thought.

Churchland says he wrote his book in part because "research into neural networks ... has produced the beginnings of a real understanding of how the biological brain works, a real understanding, that is, of how *you* work [*sic*].... [T]his idea may be found threatening, as if your innermost secrets were about to be laid bare or made public" (3). Maybe that *is* what they're feeling threatened by in San Diego, where the sun always shines and even the poor are moderately rich. But here in New York the weather is awful, the infrastructure is collapsing and even the moderately rich are poor: We should also worry about neural nets?

Chapter 9

Connectionism and the Problem of Systematicity: Why Smolensky's Solution Doesn't Work

Jerry Fodor and Brian McLaughlin

Introduction

In two recent papers, Paul Smolensky (1987, 1988b) responds to a challenge Jerry Fodor and Zenon Pylyshyn (Fodor and Pylyshyn, 1988) have posed for connectionist theories of cognition: to explain the existence of systematic relations between cognitive capacities without assuming that cognitive processes are causally sensitive to the constituent structure of mental representations. This challenge implies a dilemma: if connectionism can't account for systematicity, it thereby fails to provide an adequate basis for a theory of cognition; but if its account of systematicity requires mental processes that are sensitive to the constituent structure of mental representations, then the theory of cognition it offers will be, at best, an implementation architecture for a "classical" (language of thought) model. Smolensky thinks connectionists can steer between the horns of this dilemma if they avail themselves of certain kinds of distributed mental representation. In what follows, we will examine this proposal.

Our discussion has three parts. In section I, we briefly outline the phenomenon of systematicity and its classical explanation. As we will see, Smolensky actually offers two alternatives to this classical treatment, corresponding to two ways in which complex mental representations can be distributed; the first kind of distribution yields complex mental representations with "weak compositional structure," the second yields complex mental representations with "strong compositional structure." We will consider these two notions of distribution in turn: in section II, we argue that Smolensky's proposal that complex mental representations have weak compositional structure should be rejected as inadequate to explain systematicity and also on internal grounds; in section III, we argue that postulating mental representations with strong compositional structure also fails to provide for an explanation of systematicity. The upshot will be that Smolensky avoids only one horn of the dilemma that Fodor and Pylyshyn proposed. We shall see that his architecture is genuinely nonclassical since the representations he postulates are not "distributed over" constituents in the sense that classical representations are; and we shall see that for that very reason Smolensky's architecture leaves systematicity unexplained.

I The Systematicity Problem and Its Classical Solution

The systematicity problem is that cognitive capacities come in clumps. For example, it appears that there are families of semantically related mental states such that, as a matter of psychological law, an organism is able to be in one of the states belonging to the family only if it is able to be in many of the others. Thus you don't find organisms that can learn to prefer the green triangle to the red square but can't learn to prefer the red triangle to the green square. You don't find organisms that can think the thought that the girl loves John but can't think the thought that John loves the girl. You don't find organisms that can infer *P* from *P & Q & R* but can't infer *P* from *P & Q*. And so on over a very wide range of cases. For the purposes of this paper, we assume without argument:

> (i)   that cognitive capacities are generally systematic in the above sense, both in humans and in many infrahuman organisms;

> (ii)   that it is nomologically necessary (hence counterfactual supporting) that this is so;

> (iii)   that there must therefore be some psychological mechanism in virtue of the functioning of which cognitive capacities are systematic; and

> (iv)   that an adequate theory of cognitive architecture should exhibit this mechanism.

Any of (i)–(iv) may be viewed as tendentious; but, so far as we can tell, all four are accepted by Smolensky. So we will take them to be common ground in what follows.

The classical account of the mechanism of systematicity depends crucially on the idea that mental representation is language-like. In particular, mental representations have a combinatorial syntax and semantics. We turn to a brief discussion of the classical picture of the syntax and semantics of mental representations; this provides the basis for understanding the classical treatment of systematicity.[1]

Classical Syntax and Classical Constituents
The Classical view holds that the syntax of mental representations is like the syntax of natural language sentences in the following respect: both include complex symbols (bracketing trees) which are constructed out of what we will call classical constituents. Thus, for example, the English

sentence "John loves the girl" is a complex symbol whose decomposition into classical constituents is exhibited by some such bracketing tree as:

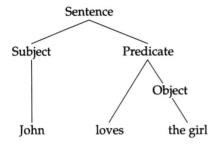

Correspondingly, it is assumed that the mental representation that is entertained when one thinks the thought that John loves the girl is a complex symbol of which the classical constituents include representations of John, the girl, and loving.

It will become clear in section III that it is a major issue whether the sort of complex mental representations that are postulated in Smolensky's theory have constituent structure. We do not wish to see this issue degenerate into a terminological wrangle. We therefore stipulate that, for a pair of expression types E1 and E2, the first is a classical constituent of the second only if the first is tokened whenever the second is tokened. For example, the English word "John" is a classical constituent of the English sentence "John loves the girl," so every tokening of the latter implies a tokening of the former (specifically, every token of the latter contains a token of the former; you can't say "John loves the girl" without saying "John").[2] Likewise, it is assumed that a Mentalese symbol which names John is a classical constituent of the Mentalese symbol that means that John loves the girl. So again a tokening of the one symbol requires a tokening of the other. It is precisely because classical constituents have this property that they are always accessible to operations that are defined over the complex symbols that contain them; in particular, it is precisely because classical mental representations have classical constituents that they provide domains for structure-sensitive mental processes. We shall see presently that what Smolensky offers as the "constituents" of connectionist mental representations are nonclassical in this respect, and that that is why his theory provides no account of systematicity.

Classical Semantics
It is part of the classical picture, both for mental representation and for representation in natural languages, that generally when a complex formula (e.g., a sentence) S expresses the proposition P, S's constituents express (or refer to) the elements of P.[3] For example, the proposition that

John loves the girl contains as its elements the individuals John and the girl, and the two-place relation loving. Correspondingly, the formula "John loves the girl," which English uses to express this proposition, contains as constituents the expressions "John," "loves," and "the girl." The sentence "John left and the girl wept," whose constituents include the formulas "John left" and "the girl wept," expresses the proposition that John left and the girl wept, whose elements include the proposition that John left and the proposition that the girl wept. And so on.

These assumptions about the syntax and semantics of mental representations are summarized by condition (C):

> (C) If a proposition P can be expressed in a system of mental representation M, then M contains some complex mental representation (a "mental sentence") S, such that S expresses P and the (classical) constituents of S express (or refer to) the elements of P.

Systematicity
The classical explanation of systematicity assumes that (C) holds by nomological necessity; it expresses a psychological law that subsumes all systematic minds. It should be clear why systematicity is readily explicable on the assumptions, first, that mental representations satisfy (C), and, second, that mental processes have access to the constituent structure of mental representations. Thus, for example, since (C) implies that anyone who can represent a proposition can ipso facto represent its elements, it implies, in particular, that anyone who can represent the proposition that John loves the girl can ipso facto represent John, the girl, and the two-place relation loving. Notice, however, that the proposition that the girl loves John is also constituted by these same individuals and relations. So, then, assuming that the processes that integrate the mental representations that express propositions have access to their constituents, it follows that anyone who can represent John's loving the girl can also represent the girl's loving John. Similarly, suppose that the constituents of the mental representation that gets tokened when one thinks that $P \& Q \& R$ and the constituents of the mental representation that gets tokened when one thinks that $P \& Q$ both include the mental representation that gets tokened when one thinks that P. And suppose that the mental processes that mediate the drawing of inferences have access to the constituent structure of mental representations. Then it should be no surprise that anyone who can infer P from $P \& Q \& R$ can likewise infer P from $P \& Q$.

To summarize: the classical solution to the systematicity problem entails that (i) systems of mental representation satisfy (C) (a fortiori, complex mental representations have classical constituents); and (ii) mental processes are sensitive to the constituent structure of mental representations. We can now say quite succinctly what our claim against Smolensky will be: on the one hand, the cognitive architecture he endorses does not provide for mental representations with classical constituents; on the other hand, he provides no suggestion as to how mental processes could be structure-sensitive unless mental representations have classical constituents; and, on the third hand (as it were) he provides no suggestion as to how minds could be systematic if mental processes aren't structure-sensitive. So his reply to Fodor and Pylyshyn fails. Most of the rest of the chapter will be devoted to making this analysis stick.

II Weak Compositionality

Smolensky's views about "weak" compositional structure are largely inexplicit and must be extrapolated from his "coffee story," which he tells in both of the papers under discussion (and also in 1988a). We turn now to considering this story.

Smolensky begins by asking how we are to understand the relation between the mental representation COFFEE and the mental representation CUP WITH COFFEE.[4] His answer to this question has four aspects that are of present interest:

(i)   COFFEE and CUP WITH COFFEE are activity vectors (according to Smolensky's weak compositional account, this is true of the mental representations corresponding to all commonsense concepts; whether it also holds for, e.g., technical concepts won't matter for what follows). A vector is, of course, a magnitude with a certain direction. A pattern of activity over a group of "units" is a state consisting of the members of the group each having an activation value of 1 or 0.[5] Activity vectors are representations of such patterns of activity.

(ii)   CUP WITH COFFEE representations contain COFFEE representations as (nonclassical) constituents in the following sense: they contain them as component vectors. By stipulation, a is a component vector of b if there is a vector x such that $a + x = b$ (where "+" is the operation of vector addition). More generally, according to Smolensky, the relation between vectors and their nonclassical constituents is that the former are derivable from the latter by operations of vector analysis.[6]

(iii)   COFFEE representations and CUP WITH COFFEE representations are activity vectors over units which represent microfeatures (units like BROWN, LIQUID, MADE OF PORCELAIN, etc.).

(iv)   COFFEE (and, presumably, any other representation vector) is context-dependent. In particular, the activity vector that is the COFFEE representation in CUP WITH COFFEE is distinct from the activity vector that is the COFFEE representation in, as it might be, GLASS WITH COFFEE or CAN WITH COFFEE. Presumably this means that the vector in question, with no context specified, does not give necessary conditions for being coffee. (We shall see later that Smolensky apparently holds that it doesn't specify sufficient conditions for being coffee either.)

Claims (i) and (ii) introduce the ideas that mental representations are activity vectors and that they have (nonclassical) constituents. These ideas are neutral with respect to the distinction between strong and weak compositionality so we propose to postpone discussing them until section III. Claim (iii), is, in our view, a red herring. The idea that there are micro-features is orthogonal both to the question of systematicity and to the issue of compositionality. We therefore propose to discuss it only briefly. It is claim (iv) that distinguishes the strong from the weak notion of compositional structure: a representation has weak compositional structure iff it contains context-dependent constituents. We propose to take up the question of context-dependent representation here.

We commence by reciting the coffee story (in a slightly condensed form).

Since, following Smolensky, we are assuming heuristically that units have bivalent activity levels, vectors can be represented by ordered sets of zeros (indicating that a unit is "off") and ones (indicating that a unit is "on"). Thus, Smolensky says, the CUP WITH COFFEE representation might be the following activity vector over microfeatures:

1-UPRIGHT CONTAINER
1-HOT LIQUID
0-GLASS CONTACTING WOOD[7]
1-PORCELAIN CURVED SURFACE
1-BURNT ODOR
1-BROWN LIQUID CONTACTING PORCELAIN
1-PORCELAIN CURVED SURFACE
0-OBLONG SILVER OBJECT
1-FINGER-SIZED HANDLE
1-BROWN LIQUID WITH CURVED SIDES AND BOTTOM[8]

This vector, according to Smolensky, contains a COFFEE representation as a constituent. This constituent can, he claims, be derived from CUP WITH COFFEE by subtracting CUP WITHOUT COFFEE from CUP WITH COFFEE. The vector that is the remainder of this subtraction will be COFFEE.

The reader will object that this treatment presupposes that CUP WITHOUT COFFEE is a constituent of CUP WITH COFFEE. Quite so. Smolensky is explicit in claiming that "the pattern or vector representing cup with coffee is composed of a vector that can be identified as a particular distributed representation of cup without coffee with a representation with the content coffee" (1988b, 10).

One is inclined to think that this must surely be wrong. If you combine a representation with the content *cup without coffee* with a representation with the content *coffee,* you get not a representation with the content *cup with coffee* but rather a representation with the self-contradictory content *cup without coffee with coffee.* Smolensky's subtraction procedure appears to confuse the representation of *cup without coffee* (viz., CUP WITHOUT COFFEE) with the representation of *cup* without the representation of *coffee* (viz., CUP). CUP WITHOUT COFFEE expresses the content *cup without coffee;* CUP combines consistently with COFFEE. But nothing does both.

On the other hand, it must be remembered that Smolensky's mental representations are advertised as context dependent, hence noncompositional. Indeed, we are given no clue at all about what sorts of relations his theory acknowledges between the semantic properties of complex symbols and the semantic properties of their constituents. Perhaps in a semantics where constituents don't contribute their contents to the symbols they belong to, it's alright after all if CUP WITH COFFEE has CUP WITHOUT COFFEE (or, for that matter, PRIME NUMBER, or GRAND-MOTHER, or FLYING SAUCER, or THE LAST OF THE MOHICANS) among its constituents.

In any event, to complete the story, Smolensky gives the following features for CUP WITHOUT COFFEE:

1-UPRIGHT CONTAINER
0-HOT LIQUID
0-GLASS CONTACTING WOOD
1-PORCELAIN CURVED SURFACE
0-BURNT ODOR
0-BROWN LIQUID CONTACTING PORCELAIN
1-PORCELAIN CURVED SURFACE
0-OBLONG SILVER OBJECT
1-FINGER-SIZED HANDLE
0-BROWN LIQUID WITH CURVED SIDES AND BOTTOM, etc.

Subtracting this vector from CUP WITH COFFEE, we get the following COFFEE representation:

0-UPRIGHT CONTAINER
1-HOT LIQUID
0-GLASS CONTACTING WOOD
0-PORCELAIN CURVED SURFACE
1-BURNT ODOR
1-BROWN LIQUID CONTACTING PORCELAIN
0-PORCELAIN CURVED SURFACE
0-OBLONG SILVER OBJECT
0-FINGER-SIZED HANDLE
1-BROWN LIQUID WITH CURVED SIDES AND BOTTOM

That, then, is Smolensky's "coffee story."

Comments

Microfeatures

It's common ground in this discussion that the explanation of system-aticity must somehow appeal to relations between complex mental repre-sentations and their constituents (on Smolensky's view, to combinatorial relations between vectors). The issue about whether there are micro-features is entirely orthogonal; it concerns only the question of which properties are expressed by the activation states of individual units. (To put it in more classical terms, it concerns the question of which symbols constitute the primitive vocabulary of the system of mental representa-tions.) If there are microfeatures, then the activation states of individual units are constrained to express only (as it might be) "sensory" properties (1987, 146). If there aren't, then activation states of individual units can express not only such properties as being brown and being hot, but also such properties as being coffee. It should be evident upon even casual reflection that, whichever way this issue is settled, the constituency ques-tion (viz., the question of how the representation COFFEE relates to the rep resentation CUP WITH COFFEE remains wide open. We therefore propose to drop the discussion of microfeatures in what follows.

Context-dependent Representation

As far as we can tell, Smolensky holds that the representation of coffee that he derives by subtraction from CUP WITH COFFEE is context dependent in the sense that it need bear no more than a "family resem-blance" to the vector that represents coffee in CAN WITH COFFEE, GLASS WITH COFFEE, etc. There is thus no single vector that counts as the COFFEE representation, hence no single vector that is a component of all the representations which, in a classical system, would have COFFEE as a classical constituent.

Smolensky himself apparently agrees that this is the wrong sort of constituency to account for systematicity and related phenomena. As he remarks, "a true constituent can move around and fill any of a number of different roles in different structures" (1988b, 11), and the connection between constituency and systematicity would appear to turn on this. For example, the solution to the systematicity problem mooted in section I depends exactly on the assumption that tokens of the representation type JOHN express the same content in the context ... LOVES THE GIRL as they do in the context THE GIRL LOVES ... (viz., that they pick out John, who is an element both of the proposition John loves the girl and of the proposition the girl loves John). It thus appears, prima facie, that the explanation of systematicity requires context-independent constituents.

How, then, does the assumption that mental representations have weak compositional structure (that is, that mental representation is context dependent) bear on the explanation of systematicity? Smolensky simply doesn't say. And we don't have a clue. In fact, having introduced the notion of weak compositional structure, Smolensky to all intents and purposes drops it in favor of the notion of strong compositional structure, and the discussion of systematicity is carried out entirely in terms of the latter. What he takes the relation between weak and strong compositional structure to be, and for that matter, which kind of structure he thinks that mental representations actually have is thoroughly unclear.[9]

In fact, quite independent of its bearing on systematicity, the notion of weak compositional structure as Smolensky presents it is of dubious coherence. We close this section with a remark or two about this point.

It looks as though Smolensky holds that the COFFEE vector that you get by subtraction from CUP WITH COFFEE is not a COFFEE representation when it stands alone. "This representation is indeed a representation of coffee, but [only?] in a very particular context: the context provided by cup [i.e., CUP]" (1987, 147). If this is the view, it has bizarre consequences. Take a liquid that has the properties specified by the microfeatures that comprise COFFEE in isolation—but that isn't coffee. Pour it into a cup, et voila! it becomes coffee by semantical magic.

Smolensky would clearly deny that the vector COFFEE that you get from CUP WITH COFFEE gives necessary conditions for being coffee, since you'd get a different COFFEE vector by subtraction from, say, GLASS WITH COFFEE. And the passage just quoted suggests that he thinks it doesn't give sufficient conditions either.[10] But, then, if the microfeatures associated with COFFEE are neither necessary nor sufficient for being coffee the question arises what, according to this story, does make a vector a COFFEE representation; when does a vector have the content *coffee*?

As far as we can tell, Smolensky holds that what makes the COFFEE component of CUP WITH COFFEE a representation with the content *coffee* is that it is distributed over units representing certain microfeatures and that it figures as a component vector of a vector that is a CUP WITH COFFEE representation. As remarked above, we are given no details at all about this reverse compositionality according to which the embedding vector determines the contents of its constituents; how it is supposed to work isn't even discussed in Smolensky's papers. But, in any event, a regress threatens since the question now arises: if being a component of a CUP OF COFFEE representation is required to make a vector a COFFEE representation, what is required to make a vector a CUP OF COFFEE representation? Well, presumably CUP OF COFFEE represents a cup of coffee because it involves the microfeatures it does and because it is a component of still another vector; perhaps one that is a THERE IS A CUP OF COFFEE ON THE TABLE representation. Does this go on forever? If it doesn't, then presumably there are some vectors that aren't constituents of any others. But now, what determines *their* contents? Not the contents of their constituents, because by assumption, Smolensky's semantics isn't compositional (CUP WITHOUT COFFEE is a constituent of CUP WITH COFFEE, etc.). And not the vectors that they are constituents of, because by assumption, there aren't any of those.

We think it is unclear whether Smolensky has a coherent story about how a system of representations could have weak compositional structure.

What, in light of all this, leads Smolensky to embrace his account of weak compositionality? Here's one suggestion: perhaps Smolensky confuses being a representation of a cup with coffee with being a CUP WITH COFFEE representation. Espying some cup with coffee on a particular occasion, in a particular context, one might come to be in a mental state that represents it as having roughly the microfeatures that Smolensky lists. That mental state would then be a representation of a cup with coffee in this sense: there is a cup of coffee that it's a mental representation of. But it wouldn't, of course, follow that it's a CUP WITH COFFEE representation; and the mental representation of that cup with coffee might be quite different from the mental representation of the cup with coffee that you espied on some other occasion or in some other context. So which mental representation a cup of coffee gets is context dependent, just as Smolensky says. But that doesn't give Smolensky what he needs to make mental representations themselves context dependent. In particular, from the fact that cups with coffee get different representations in different contexts, it patently doesn't follow that the mental symbol that represents something as being a cup of coffee in one context might represent something as being something else (a giraffe say, or the last of the Mohicans) in some other context. We doubt that anything will give Smolensky that, since we know of no reason to suppose that it is true.

In short, it is natural to confuse the true but uninteresting thought that how you mentally represent some coffee depends on the context, with the much more tendentious thought that the mental representation COFFEE is context dependent. Assuming that he is a victim of this confusion makes sense of many of the puzzling things Smolensky says in the coffee story. Notice, for example, that all the microfeatures in his examples express more or less perceptual properties (cf. Smolensky's own remark that his microfeatures yield a "nearly sensory level representation.") Notice, too, the peculiarity that the microfeature "porcelain curved surface" occurs twice in the vectors for CUP WITH COFFEE, COFFEE, CUP WITHOUT COFFEE and the like. Presumably, what Smolensky has in mind is that when you look at a cup, you see two curved surfaces, one going off to the left and the other going off to the right.

Though we suspect this really is what's going on, we won't pursue this interpretation further, since if it's correct, the coffee story is completely irrelevant to the question of what kind of constituency relation a COFFEE representation has to a CUP WITH COFFEE representation; and that, remember, is the question that bears on the issues about systematicity.

III Strong Compositional Structure

So much, then, for "weak" compositional structure. Let us turn to Smolensky's account of "strong" compositional structure. Smolensky says that

> [a] true constituent can move around and fill any of a number of different roles in different structures. Can this be done with vectors encoding distributed representations, and be done in a way that doesn't amount to simply implementing symbolic syntactic constituency? The purpose of this section is to describe research showing that the answer is affirmative. (1988b, 11)

The idea that mental representations are activity vectors over units, and the idea that some mental representations have other mental representations as components, is common to the treatment of both weak and strong compositional structure. However, Smolensky's discussion of the latter idea differs in several respects from his discussion of the former. First, he explicitly supposes that units have continuous activation levels between 0 and 1; second, he does not invoke the idea of microfeatures when discussing strong compositional structure; third, he introduces a new vector operation (multiplication) to the two previously mentioned (addition and subtraction); fourth, and most important, on his view strong compositional structure does not invoke—indeed, it would appear to be incompatible with—the notion that mental representations are context dependent. So strong compositional structure does not exhibit the incoherence of Smolensky's theory of context-dependent representation.

We will proceed as follows. First we briefly present the notion of strong compositional structure. Then we turn to criticism.

Smolensky explains the notion of strong compositional structure, in part, by appeal to the ideas of a tensor product representation and a superposition representation. To illustrate these ideas, consider how a connectionist machine might represent four-letter English words. Words can be decomposed into roles (viz., ordinal positions that letters can occupy) and things that can fill these roles (viz., letters). Correspondingly, the machine might contain activity vectors over units that represent the relevant roles (i.e., over the role units) and activity vectors over units that represent the fillers (i.e., over the filler units). Finally, it might contain activity vectors over units that represent filled roles (i.e., letters in letter positions); these are the binding units. The key idea is that the activity vectors over the binding units might be tensor products of activity vectors over the role units and the filler units. The representation of a word would then be a superposition vector over the binding units; that is, a vector that is arrived at by superimposing the tensor product vectors.

The two operations used here to derive complex vectors from component vectors are vector multiplication (in the case of tensor product vectors) and vector addition (in the case of superposition vectors). These are iterative operations in the sense that activity vectors that result from the multiplication of role vectors and filler vectors might themselves represent the fillers of roles in more complex structures. Thus a tensor product that represents the word "John" as "J" in first position, "o" in second position, etc., might itself be bound to the representation of a syntactical function to indicate, for example, that "John" has the role subject-of in "John loves the girl." Such tensor product representations could themselves be superimposed over yet another group of binding units to yield a superposition vector that represents the bracketing tree (John) (loves (the girl)).[11]

It is, in fact, unclear whether this sort of apparatus is adequate to represent all the semantically relevant syntactic relations that classical theories express by using bracketing trees with classical constituents. (There are, for example, problems about long-distance binding relations, as between quantifiers and bound variables.) But we do not wish to press this point. For present polemical purposes, we propose simply to assume that each classical bracketing tree can be coded into a complex vector in such a fashion that the constituents of the tree correspond in some regular way to components of the vector.

But to assume this is not, of course, to grant that either tensor product or superposition vectors have classical constituent structure. In particular, from the assumptions that bracketing trees have classical constituents and that bracketing trees can be coded by activity vectors, it does not follow that activity vectors have classical constituents. On the contrary, a point about which Smolensky is himself explicit is vital in this regard: the components of a complex vector need not even correspond to patterns of activity over units actually in the machine. As Smolensky puts it, the activity states of the filler and role units can be "imaginary" even though the ultimate activity vectors—the ones that do not themselves serve as filler or role components of more complex structures—must be actual activity patterns over units in the machine. Consider again our machine for representing four-letter words: the superposition pattern that represents, say, the word "John" will be an activity vector actually realized in the machine. However, the activity vector representing "J" will be merely imaginary, as will the activity vector representing the first letter position. Similarly for the tensor product activity vector representing "J" in the first letter position. The only pattern of activity that will actually be tokened in the machine is the superposition vector representing "John."

These considerations are of central importance for the following reason. Smolensky's main strategy is, in effect, to invite us to consider the components of tensor product and superposition vectors to be analogous to the classical constituents of a complex symbol, and hence to view them as providing a means by which connectionist architectures can capture the causal and semantic consequences of classical constituency in mental representations. However, the components of tensor product and superposition vectors differ from classical constituents in the following way: when a complex classical symbol is tokened, its constituents are tokened. When a tensor product vector or superposition vector is tokened, its components are not (except per accidens). The implication of this difference, from the point of view of the theory of mental processes, is that whereas the classical constituents of a complex symbol are, ipso facto, available to contribute to the causal consequences of its tokening—in particular, they are available to provide domains for mental processes—the components of tensor product and superposition vectors can have no causal status as such. What is merely imaginary can't make things happen, to put this point in a nutshell.

We will return presently to what these differences imply for the treatment of the systematicity problem. There is, however, a preliminary issue that needs to be discussed.

We have seen that the components of tensor product and superposition vectors, unlike classical constituents, are not, in general, tokened whenever the activity vector of which they are the components is tokened. It is worth emphasizing, in addition, the familiar point that there is in general no unique decomposition of a tensor product or superposition vector into components. Indeed, given that units are assumed to have continuous levels of activation, there will be infinitely many decompositions of a given activity vector. One might wonder, then, what sense there is in talk of the decomposition of a mental representation into significant constituents, given the notion of constituency that Smolensky's theory provides.

Smolensky replies to this point as follows. Cognitive systems will be dynamical systems; there will be dynamic equations over the activation values of individual units, and these will determine certain regularities over activity vectors. Given the dynamical equations of the system, certain decompositions can be especially useful for "explaining and understanding" the system's behavior. In this sense, the dynamics of a system may determine "normal modes" of decomposition into components. So, for example, though a given superposition vector in principle can be taken to be the sum of many different sets of vectors, it may yet turn out that we get a small group of sets—even a unique set—when we decompose in the direction of normal modes; and likewise for decomposing tensor product vectors. The long and short is that it could, in principle, turn out that given the (thus far undefined) normal modes of a dynamical cognitive system, complex superposition vectors will have it in common with classical complex symbols that they have a unique decomposition into semantically significant parts. Of course, it also could turn out that they don't, and no ground for optimism on this point has thus far been supplied.

Having noted this problem, however, we propose simply to ignore it. So here is where we now stand: by assumption (though quite possibly contrary to fact), tensor product vectors and superposition vectors can code constituent structure in a way that makes them adequate vehicles for the expression of propositional content; and, by assumption (though again quite possibly contrary to fact), the superposition vectors that cognitive theories acknowledge have a unique decomposition into semantically interpretable tensor product vectors, which in turn have a unique decomposition into semantically interpretable filler vectors and role vectors; so it's determinate which proposition a given complex activity vector represents.

Now, assuming all this, what about the systematicity problem?

The first point to make is this: if tensor product or superposition vector representation solves the systematicity problem, the solution must be quite different from the classical proposal sketched in section I. True, tensor product vectors and superposition vectors "have constituents" in some suitably extended sense: tensor product vectors have semantically evaluable components, and superposition vectors are decomposable into semantically evaluable tensor product vectors. But the classical solution to the systematicity problem assumes that the constituents of mental representations have causal roles; that they provide domains for mental processes. The classical constituents of a complex symbol thus contribute to determining the causal consequences of tokening that symbol, and it seems clear that the "extended" constituents of a tensor product or a superposition representation can't do that. On the contrary, the components of a complex vector are typically not even tokened when the complex vector itself is tokened; they are simply constituents into which the complex vector *could* be resolved consonant with decomposition in the direction of normal modes. But, to put it crudely, the fact that six could be represented as "3 × 2" cannot, in and of itself, affect the causal processes in a computer (or a brain) in which six is represented as "6." Merely counterfactual representations have no causal consequences; only actually tokened representations do.

Smolensky is, of course, sensitive to the question whether activity vectors really do have constituent structure. He defends at length the claim that he has not distorted the notion of constituency in saying that they do. Part of this defense adverts to the role that tensor products and superpositions play in physical theory:

> The state of the atom, like the states of all systems in quantum theory, is represented by a vector in an abstract vector space. Each electron has an internal state (its "spin"); it also has a role it plays in the atom as a whole: it occupies some "orbital," essentially a cloud of probability for finding it at particular places in the atom. The internal state of an electron is represented by a "spin vector"; the orbital or role of the electron (part) in the atom (whole) is represented by another vector, which describes the probability cloud. The vector representing the electron as situated in the atom is the tensor product of the vector representing the internal state of the electron and the vector representing its orbital. The atom as a whole is represented by a vector that is the sum or superposition of vectors, each of which represents a particular electron in its orbital.... (1988b, 19–20)

"So," Smolensky adds, "someone who claims that the tensor product representational scheme distorts the notion of constituency has some explaining to do" (1988b, 20).

The physics lesson is greatly appreciated; but it is important to be clear on just what it is supposed to show. It's not, at least for present purposes, in doubt that tensor products can represent constituent structure. The relevant question is whether tensor product representations *have* constituent structure; or, since we have agreed that they may be said to have constituent structure "in an extended sense," it's whether they have the kind of constituent structure to which causal processes can be sensitive, hence the kind of constituent structure to which an explanation of systematicity might appeal. But we have already seen the answer to this question: the constituents of complex activity vectors typically aren't "there," so if the causal consequences of tokening a complex vector are sensitive to its constituent structure, that's a miracle.[12]

We conclude that assuming that mental representations are activation vectors does not allow Smolensky to endorse the classical solution of the systematicity problem. And, indeed, we think Smolensky would grant this since he admits up front that mental processes will not be causally sensitive to the strong compositional structure of mental representations. That is, he acknowledges that the constituents of complex mental representations play no causal role in determining what happens when the representations get tokened: "Causal efficacy was not my goal in developing the tensor product representation" (1988b, 21). What are causally efficacious according to connectionists are the activation values of individual units; the dynamical equations that govern the evolution of the system will be defined over these. It would thus appear that Smolensky must have some nonclassical solution to the systematicity problem up his sleeve, some solution that does not depend on assuming mental processes that are causally sensitive to constituent structure. So then, after all this, what is Smolensky's solution to the systematicity problem?

Remarkably enough, Smolensky doesn't say. All he does say is that he "hypothesizes ... that ... the systematic effects observed in the processing of mental representations arise because the evolution of vectors can be (at least partially and approximately) explained in terms of the evolution of their components, even though the precise dynamical equations apply [only] to the individual numbers comprising the vectors and [not] at the level of [their] constituents—i.e. even though the constituents are not causally efficacious" (1988b, 21).

It is left unclear how the constituents ("components") of complex vectors are to explain their evolution (even partially and approximately) when they are, by assumption, at best causally inert, and at worst merely imaginary. In any event, what Smolensky clearly does think is causally responsible for the "evolution of vectors" (and hence for the systematicity of cognition) are unspecified processes that affect the states of activation of the individual units (the neuron analogs) out of which the vectors are composed. So, then, as far as we can tell, the proposed connectionist explanation of systematicity (and related features of cognition) comes down to this: Smolensky "hypothesizes" that systematicity is somehow a consequence of underlying neural processes.[13] Needless to say, if that is

Smolensky's theory, it is on the one hand certainly true, and on the other hand not intimately dependent upon his long story about fillers, binders, tensor products, superposition vectors, and the rest.

By way of rounding out the argument, we want to reply to a question raised by an anonymous *Cognition* reviewer, who asks: "couldn't Smolensky easily build in mechanisms to accomplish the matrix algebra operations that would make the necessary vector explicit (or better yet, from his point of view, ... mechanisms that are sensitive to the imaginary components without literally making them explicit in some string of units)?"[14] But this misses the point of the problem that systematicity poses for connectionists, which is not to show that systematic cognitive capacities are *possible* given the assumptions of a connectionist architecture, but to explain how systematicity could be *necessary*—how it could be a law that cognitive capacities are systematic—given those assumptions.[15]

No doubt it is possible for Smolensky to wire a network so that it supports a vector that represents *aRb* if and only if it supports a vector that represents *bRa*; and perhaps it is possible for him to do that without making the imaginary units explicit (though there is, so far, no proposal about how to ensure this for arbitrary *a*, *R*, and *b*). The trouble is that, although the architecture permits this, it equally permits Smolensky to wire a network so that it supports a vector that represents *aRb* if and only if it supports a vector that represents *zSq*; or, for that matter, if and only if it supports a vector that represents the last of the Mohicans. The architecture would appear to be absolutely indifferent between these options.

In contrast, as we keep saying, in the classical architecture, if you meet the conditions for being able to represent *aRb*, *you cannot but meet the conditions for being able to represent bRa*; the architecture won't let you do so because (i) the representation of *a*, *R*, and *b* are constituents of the representation of *aRb*, and (ii) you have to token the constituents of the representations that you token, so classical constituents can't be just imaginary. So then: it is built into the classical picture that you can't think *aRb* unless you are able to think *bRa*, but the connectionist picture is neutral on whether you can think *aRb* even if you can't think *bRa*. But it is a law of nature that you can't think *aRb* if you can't think *bRa*. So the classical picture explains systematicity and the connectionist picture doesn't. So the classical picture wins.

Conclusion

At one point in his discussion, Smolensky makes some remarks that we find quite revealing: he says that even in cases that are paradigms of classical architectures (LISP machines and the like), "we normally think of the 'real' causes as physical and far below the symbolic level: ..." Hence, even in classical machines, the sense in which operations at the symbol level are real causes is just that "there is ... a complete and precise algorithmic (temporal) story to tell about the states of the machine described" at that level (1988b, 20). Smolensky, of course, denies that there is a "comparable story at the symbolic level in the human cognitive architecture ... that is a difference with the classical view that I have made much of. *It may be that a good way to characterize the difference is in terms of whether the constituents in mental structure are causally efficacious in mental processing*" (1988b, 20; our emphasis).

We say that this comment is revealing because it suggests a diagnosis: it would seem that Smolensky has succumbed to a sort of generalized epiphenomenalism. The idea is that even classical constituents participate in causal processes solely by virtue of their physical microstructure, so that even on the classical story it's what happens at the neural level that really counts. Though the evolution of vectors can perhaps be explained in a predictively adequate sort of way by appeal to macroprocesses like operations on constituents, still, if you want to know what's really going on—if you want the causal explanation—you need to go down to the "precise dynamical equations" that apply to activation states of units. That intentional generalizations can only approximate these precise dynamical equations is among Smolensky's recurrent themes. By conflating the issue about "precision" with the issue about causal efficacy, Smolensky makes it seem that to the extent that macrolevel generalizations are imprecise, macrolevel processes are epiphenomenal.

It would need a philosophy lesson to say all of what's wrong with this view. Suffice it for present purposes that his argument iterates in a way that Smolensky ought to find embarrassing. No doubt, we do get greater precision when we go from generalizations about operations on constituents to generalizations about operations on units. But if that shows that symbol-level processes aren't really causal, then it must be that unit-level processes aren't really causal either. After all, we get still more precision when we go down from unit-sensitive operations to molecule-sensitive operations, and more precision yet when we go down from molecule-sensitive operations to quark-sensitive operations. The moral is not, however, that the causal laws of psychology should be stated in terms of the behavior of quarks. Rather, the moral is that whether you have a level of causal explanation is a question, not just of how much precision you are able to achieve, but also of what generalizations you are able to express. The price you pay for doing psychology at the level of units is that you lose causal generalizations that symbol-level theories are able to state. Smolensky's problems with capturing the generalizations about systematicity provide a graphic illustration of these truths.

It turns out, at any event, that there is a crucial caveat to Smolensky's repeated claim that connectionist mechanisms can reconstruct everything that's interesting about the notion of constituency. Strictly speaking, he claims only to reconstruct whatever is interesting about constituents except their causes and effects. The explanation of systematicity turns on the causal role of the constituents of mental representations and is therefore among the casualties. Hilary Putnam, back in the days when he was still a metaphysical realist, used to tell a joke about a physicist who actually managed to build a perpetual motion machine; all except for a part that goes back and forth, back and forth, back and forth, forever. Smolensky's explanation of systematicity has very much the character of this machine.

We conclude that Fodor and Pylyshyn's challenge to connectionism has yet to be met. We still don't have even a suggestion of how to account for systematicity within the assumptions of connectionist cognitive architecture.[16]

Notes

1. Since the two are often confused, we wish to emphasize that taking systematicity for granted leaves the question of compositionality wide open. The systematicity of cognition consists in, for example, the fact that organisms that can think aRb can think bRa and vice versa. Compositionality proposes a certain explanation of systematicity, viz., that the content of thoughts is determined in a uniform way by the content of the context-independent concepts that are their constituents, and that the thought that bRa is constituted by the same concepts as the thought that aRb. So the polemical situation is as follows. If you are a connectionist who accepts systematicity, then you must argue either that systematicity can be explained without compositionality, or that connectionist architecture accommodates compositional representation. So far as we can tell, Smolensky vacillates between these options; what he calls "weak compositionality" favors the former and what he calls "strong compositionality" favors the latter.

   We emphasize this distinction between systematicity and compositionality in light of some remarks by an anonymous Cognition reviewer: "By berating the [connectionist] modelers for their inability to represent the common-sense [uncontextualized] notion of 'coffee' ... Fodor and McLaughlin are missing a key point—the models are not supposed to do so. If you buy the ... massive context-sensitivity ... that connectionists believe in." Our strategy is not, however, to argue that there is something wrong with connectionism because it fails to offer an uncontextualized notion of mental (or, mutatis mutandis, linguistic) representation. Our argument is that if connectionists assume that mental representations are context sensitive, they will need to offer some explanation of systematicity that does not entail compositionality; and they do not have one.

   We don't, therefore, offer direct arguments for context-insensitive concepts in what follows; we are quite prepared that "coffee" should have a meaning only in context. Only, we argue, if it does, then some noncompositional account of the systematicity of coffee-thoughts will have to be provided.

2. Though we shall generally consider examples where complex symbols literally contain their classical constituents, the present condition means to leave it open that symbols may have classical constituents that are not among their (spatio-temporal) parts.

(For example, so far as this condition is concerned, it might be that the classical constituents of a symbol include the values of a "fetch" operation that takes the symbol as an argument.)

3. We assume that the elements of propositions can include, for example, individuals, properties, relations, and other propositions. Other metaphysical assumptions are, of course, possible. For example, it is arguable that the constituents of propositions include individual concepts (in the Fregean sense) rather than individuals themselves; and so on. Fortunately, it is not necessary to enter into these abstruse issues to make the points that are relevant to the systematicity problem. All we really need is that propositions have internal structure, and that, characteristically, the internal structure of complex mental representations corresponds in the appropriate way to the internal structure of the propositions that they express.

4. The following notational conventions will facilitate the discussion: we will follow standard practice and use capitalized English words and sentences as canonical names for mental representations. (Smolensky uses italicized English expressions instead.) We stipulate that the semantic value of a mental representation so named is the semantic value of the corresponding English word or sentence, and we will italicize words or sentences that denote semantic values. So, for example, COFFEE is a mental representation that expresses (the property of) *being coffee* (as does the English word "coffee"); JOHN LOVES THE GIRL is a mental representation that expresses the proposition that *John loves the girl*; and so forth. It is important to notice that our notation allows that the mental representation JOHN LOVES THE GIRL can be atomic and the mental representation COFFEE can be a complex symbol. That is, capitalized expressions should be read as the names of mental representations rather than as their structural descriptions.

5. Smolensky apparently allows that units may have continuous levels of activation from 0 to 1. In telling the coffee story, however, he generally assumes bivalence for ease of exposition.

6. As we shall see below, when an activity vector is tokened, its component vectors typically are not. So the constituents of a complex vector are, ipso facto, nonclassical.

7. Notice that this microfeature is "off" in CUP WITH COFFEE, so it might be wondered why Smolensky mentions it at all. The explanation may be this: operations of vector combination apply only to vectors of the same dimensionality. In the context of the weak constituency story, this means that you can only combine vectors that are activity patterns over the same units. It follows that a component vector must contain the same units (though possibly at different levels of activation) as the vectors with which it combines. Thus if GRANNY combines with COFFEE to yield GRANNY'S COFFEE, GRANNY must contain activation levels for all the units in COFFEE and vice versa. In the present example, it may be that CUP WITH COFFEE is required to contain a 0-activation level for GLASS CONTACTING WOOD to accommodate cases where the former is a component of some other vector. Similarly with OBLONG SILVER OBJECT (below) since cups with coffee often have spoons in them.

8. Presumably Smolensky does not take this list to be exhaustive, but we don't know how to continue it. Beyond the remark that although the microfeatures in his examples correspond to "nearly sensory-level representation[s]" that fact is "not essential," Smolensky provides no account at all of what determines which contents are expressed by microfeatures. The question thus arises why Smolensky assumes that COFFEE is not itself a microfeature. In any event, Smolensky repeatedly warns the reader not to take his examples of microfeatures very seriously, and we don't.

9. They can't have both; either the content of a representation is context dependent or it's not. So, if Smolensky does think that you need strong compositional structure to explain systematicity, and that weak compositional structure is the kind that connectionist

representations have, then it would seem that he thereby grants Fodor and Pylyshyn's claim that connectionist representations can't explain systematicity. We find this all very mysterious.

10. If they were necessary and sufficient, COFFEE wouldn't be context dependent.

11. The function of the brackets in a classical bracketing tree is precisely to exhibit its decomposition into constituents; and when the tree is well formed this decomposition will be unique. Thus the bracketing of "(John) (loves) (the girl)" implies, for example, both that "the girl" is a constituent and that "loves the" is not.

12. It's a difference between psychology and physics that whereas psychology is about the casual laws that govern tokenings of (mental) representations, physics is about the causal laws that govern (not mental representations but) atoms, electrons, and the like. Since being a representation isn't a property in the domain of physical theory, the question of whether mental representations have constituent structure has no analogue in physics.

13. More precisely: we take Smolensky to be claiming that there is some property D, such that if a dynamical system has D its behavior is systematic, and such that human behavior (for example) is caused by a dynamical system that has D. The trouble is that this is a platitude, since it is untendentious that human behavior is systematic, that its causation by the nervous system is lawful, and that the nervous system is dynamical. The least that has to happen if we are to have a substantive connectionist account of systematicity is: first, it must be made clear what property D is, and second, it must be shown that D is a property that connectionist systems can have by law. Smolensky's theory does nothing to meet either of these requirements.

14. Actually, Smolensky is forced to choose the second option. To choose the first would be, in effect, to endorse the classical requirement that tokening a symbol implies tokening its constituents—in which case, the question arises once again why such a network isn't an implementation of a language-of-thought machine. Just as Smolensky mustn't allow the representations of roles, fillers, and binding units to be subvectors of superposition vectors if he is to avoid the "implementation" horn of Fodor and Pylyshyn's dilemma, so too he must avoid postulating mechanisms that make role, filler, and binding units explicit (specifically, accessible to mental operations) whenever the superposition vectors are tokened. Otherwise he again has symbols with classical constituents and raises the question of why the proposed device isn't a language-of-thought machine. Smolensky's problem is that the very feature of his representations that makes them wrong for explaining systematicity (viz., that their constituents are allowed to be imaginary) is the one that they have to have to ensure that they aren't classical.

15. Fodor and Pylyshyn were very explicit about this. See, for example, 1988, 48.

16. Terence Horgan remarks (personal communication), "... often there are two mathematically equivalent ways to calculate the time-evolution of a dynamical system. One is to apply the relevant equations directly to the numbers that are elements of a single total vector describing the initial state of the system. Another way is to mathematically decompose that vector into component normal-mode vectors, then compute the time evolution of each [of these] ... and then take the later state of the system to be described by a vector that is the superposition of the resulting normal-mode vectors." Computations of the former sort are supposed to be the model for operations that are "sensitive" to the components of a mental representation vector without recovering them. (Even in the second case, it's the theorist who recovers them in the course of the computations by which he makes his predictions. This does not, of course, imply that the constituents thus "recovered" participate in causal processes in the system under analysis.)

Chapter 10

Connectionism and the Problem of Systematicity (Continued): Why Smolensky's Solution *Still* Doesn't Work

Paul Smolensky has recently announced that the problem of explaining the compositionality of concepts within a connectionist framework is solved in principle. Mental representations are vectors over the activity states of connectionist "units," but the vectors encode classical trees, whose structural properties in turn "acausally" explain the facts of compositionality. This sounds suspiciously like the offer of a free lunch, and it turns out, upon examination, that there is nothing to it.

Human cognition exhibits a complex of closely related properties—including systematicity, productivity, and compositionality—which a theory of cognitive architecture ignores at its peril.[1] If you are stuck with a theory that denies that cognition has these properties, you are dead and gone. If you are stuck with a theory that is compatible with cognition having these properties but is unable to explain why it has them, you are, though arguably still breathing, clearly in deep trouble. There are, to be sure, cognitive scientists who do not play by these rules, but I propose to ignore them in what follows. Paul Smolensky, to his credit, is not among them.

Smolensky has recently been spending a lot of his time trying to show that, vivid first impressions to the contrary notwithstanding, some sort of connectionist cognitive architecture can indeed account for compositionality, productivity, systematicity, and the like. It turns out to be rather a long story how this is supposed to work; 185 pages of a recent collection of papers on connectionism (Macdonald and Macdonald, 1995) are devoted to Smolensky's telling of it, and there appears still to be no end in sight. It seems it takes a lot of squeezing to get this stone to bleed.

Still, Smolensky's account of compositional phenomena has had a good press in some quarters; the Churchlands tell us, for example, that "Smolensky (1990) has shown that [his sort of cognitive architecture is] at least adequate to embody ... the systematic linguistic structures and transformations deemed essential by Fodorean accounts of cognition"

(McCauley, 1995, 234). It would certainly be big news if that were true. But, in fact, Smolensky's account doesn't work, and the formal details, though they are daunting and take up a lot of pages in Smolensky's papers, are largely inessential to understanding what's gone wrong.

To begin, I want to sketch an area of consensus. Unlike the kind of cognitive theory that Smolensky calls "local" connectionism, and quite like what have come to be called "classical" or "language of thought"[2] cognitive architectures, Smolensky architectures[3] invoke a structural relation between concepts (mutatis mutandis, between mental representations)[4] to explain compositional phenomena. The accounts of compositionality that S-architectures and classical theories propose are thus of the same general form. Both turn essentially on postulating an (asymmetric) structural relation (call it R) that holds between concepts and constrains their possession conditions and their semantics. In particular, according to both S-architectures and classical architectures:

(i)   (possession conditions) If a concept C bears R to a concept C^*, then having C^* requires having C; and

(ii)   (semantics) If a concept C bears R to a concept C^*, then the content of C^* is determined (at least inter alia) by the content of C.

We'll see presently that what distinguishes S-architectures from classical architectures is *what relation they say R is.*

I can now tip my polemical hand: I'm going to argue that the account of R that S-theories give is entirely and hopelessly parasitic on the account of R that classical architectures give. That's the sense in which, even after all those pages, Smolensky *hasn't so much as made a start* on constructing an alternative to the classical account of compositional phenomena. I propose to take this in three stages: First, I want to remind you of what classical theories say about relation R. I'll then say how Smolensky's putative alternative is supposed to work. I'll then say why Smolensky's alternative is *merely* putative. Then I'll stop.

Stage 1: Classical Theories

According to classical architectures, R is the *constituency* relation: C bears R to C^* iff C is a constituent of C^*. It's worth getting clear on how the assumption that R is the constituency relation connects with assumptions (i) and (ii) and with the classical account of compositional phenomena.

The Connection with (i)
Roughly, the constituency relation is a *part/whole* relation: If C is a constituent of C^*, then a token of C is a part of every token of C^*. More pre-

cisely, constituency is a *co-tokening* relation; that is, if C is a constituent of C^*, then it is metaphysically necessary that for every tokening of C^* there is a corresponding tokening of C. (Part/whole isn't, of course, the only relation of co-tokening. For present purposes, it doesn't matter which of these you choose as R. For discussion, see chapter 9.) Since on this construal of R, nobody can *token* C^* without tokening C, it follows that nobody can *have* C^* without having C.

The Connection with (ii)
Reading R as the constituency relation doesn't literally entail (ii) since presumably it's possible for a system of representations to exhibit syntactic constituency without exhibiting semantic compositionality (in effect, all the syntactically complex symbols in such a system would be idioms).[5] But taking R to be the constituency relation does provide for a natural explication of the informal idea that mental representations are compositional: what makes them compositional is that the content of structurally complex mental symbols is inherited from the contents of their less structurally complex parts.

To claim that this is a "natural" way to explicate compositionality is to put the case very mildly. If R is constituency, then (ii) says that the semantics of complex representations derives from the semantics *of their parts*. The effect is to make the connection between compositionality, systematicity, productivity, and the like immediately clear; an example may serve to give the flavor of the thing.

There seems to be a strong pretheoretic intuition that to think the content *brown cow* is ipso facto to think the content *brown*. This amounts to considerably more than a truism; for example, it isn't explained just by the *necessity* of the inference from *brown cow* to *brown*. (Notice that the inference from *two* to *prime* is necessary too, but it is *not* intuitively plausible that you can't think *two* without thinking *prime*.) Moreover, the intuition that thinking *brown cow* somehow requires thinking *brown* is pretty clearly connected with corresponding intuitions about the systematicity of the concepts involved. It's plausibly *because* thinking the content *brown cow* somehow invokes the *concept* BROWN that anybody who is able to think *brown cow* and *tree* is ipso facto able to think *brown tree*.

Assuming that (ii) holds and that R is the constituency relation (or some other co-tokening relation; see above) is a way to guarantee that thinking *brown cow* does indeed invoke the concept BROWN, so it provides an elegant solution to the galaxy of observations I just retailed. If the mental representation BROWN is *a part of* the mental representation BROWN COW, then *of course* you can't have a token of the former in your head without having a token of the latter in your head too. Systematicity likewise falls into place: If the mental representation BROWN is a part both of the mental representation BROWN COW and of the mental representation BROWN TREE, then if you're in a position to think the intentional (semantic) content that BROWN contributes to BROWN COW, you are *thereby* in a position to think the intentional (semantic) content that BROWN contributes to BROWN TREE . Constituency and compositionality, taken together, guarantee that all of this is so.

Very satisfactory, I would have thought; so I want to emphasize that none of this follows from (ii) alone (i.e., from (ii) *without* the identification of R with constituency). (ii) says only that if C bears R to C^*, then C (partially) determines the intentional and semantic content of C^*. But that doesn't come close to entailing that you can't think C^* without thinking C; and, as we've been seeing, it's this latter constraint that the explanation of compositional phenomena appears to turn on. What accounts for compositionality, according to classical architectures, is (ii) *together with the assumption that if C bears R to C^*, then C is co-tokened with C^**; and this assumption is guaranteed, in turn, by the identification of R with constituency. (More on all this later in the discussion.)

Stage 2: Smolenksy Architectures

For present purposes, the briefest sketch will do.

According to Smolensky, mental representations are *vectors*: "[T]he import of distributed representation is precisely that ... a representation is not to be found at a unit but in an activity pattern. Mathematically, an activity pattern is ... a list of numbers ... giving the activity of all the units over which the pattern resides" (232). All you need to know about such representations, for present purposes, is that they do *not* decompose into constituents. Of course lists can have sublists, and every sublist of a vector specifies the activation level of some unit(s) or other. But there is, for example, no guarantee that the list of numbers that expresses the content *brown cow* and the list of numbers that expresses the content *brown tree* will have any of their sublists in common. (Quite generally, the fact that a certain vector has a semantic interpretation does not even guarantee that its subvectors have semantic interpretations too. This is because it's not guaranteed—in fact, it's not usually the case—that every "unit" has a semantic interpretation.) We're about to see that S-architectures do offer a relation R that makes (ii) true, but it isn't a part/whole relation; in fact, it isn't a co-tokening relation of any kind.

So, if R isn't constituency, what is it? Roughly, according to S-architectures, R is a *derivation* relation. We arrive at the core of Smolensky's theory. The core of Smolensky's theory is an algorithm for encoding constituent structure trees as vectors. Once again, the details don't matter (for a sketch, see Smolensky [1995] circa 236), but for the sake of the argument, I'll suppose (with Smolensky) that the encoding algorithm is bi-unique: Given a tree, the algorithm yields a corresponding vector; given a vector derived from a tree, the algorithm yields the tree from which the vector is derived.

Here's how things stand so far: Vectors don't have constituents. But vectors can be derived from trees; and, of course, trees *do* have constituents. Smolensky's proposal (though he doesn't put it quite this way) is that we introduce a notion of *derived constituency* for vectors that works as follows: C is a *derived constituent of* vector V iff V (uniquely) encodes C^* and C is a constituent of C^*. That is, the *derived* constituents of a vector V are the constituents tout court of the tree that V encodes. So, for example, the vector for *brown* needn't (and generally won't) be a part of the vector for *brown cow*. But if (by assumption) V encodes the tree ((brown) (cow)), then the subtree (brown) is a classical constituent of the tree that V encodes. So it follows from the definition that the subtree (brown) is a *derived* constituent of V. So, according to S-architectures, R is a sort of constituency relation after all, only it's, as it were, a constituency relation once removed: Though it's *not* a part/whole relation over vectors, it *is* a part/whole relation over what the vectors encode.

As you will probably have gathered, I don't actually care much whether this apparatus can be made to operate (e.g., whether there is an algorithm of the sort that S-architecture wants to use to encode vectors into trees and vice versa). But I am *very* interested in the question of whether, if the apparatus can be made to operate, then a connectionist architecture can explain the compositionality. So let's turn to that.

Stage 3: Why Smolensky's Solution Still Doesn't Work

To begin, I want to reiterate a point that chapter 9 insisted on. The explanations of compositional phenomena that classical theories offer are, in general, quite naturally construed as *causal* explanations. That this is so follows from the classical construal of R as constituency. Constituency is, as previously remarked, a co-tokening relation; so if BROWN is a constituent of BROWN COW, it can hardly be surprising that whenever a token of the latter is available to play a causal role in a mental process, *so too is a token of the former.* This is the heart of the classical account of compositionality: If the effects of tokening BROWN COW partially overlap with the effects of tokening BROWN TREE, that's because, in a perfectly literal sense, the causes of these effects do too.

Now, patently, this pattern of explanation is not available to an S-architecture; remember, the tokening of the BROWN COW vector does *not* guarantee a co-tokening of the BROWN vector, so it can't be taken for granted that the effects of a tokening of BROWN COW will include the effects of a tokening of BROWN. Mental representations are, by definition, necessarily co-tokenend with their *classical* constituents. But it is not the case that they are, by definition or otherwise, necessarily co-tokened with their *derived* constituents. In fact, it's perfectly possible that a vector should be tokened *even if none of its derived constituents ever was or ever will be.* Remember that *all* that connects the vector V with its derived constituents is that there is a biunique algorithm which, *if it were executed, would compute* a tree such that the *derived* constituents of the vector are the *real* constituents of the tree. But nothing in the notion of an S-architecture, or in the kind of explanations S-architectures provide, requires that such subjunctives ever get cashed. It is perfectly alright for C to be a derived constituent of vector token V, even though all the tokens of C are, and always will be, entirely counterfactual.

Now, I suppose it goes without saying that merely *counterfactual* causes don't have any actual effects. So explanations that invoke the *derived* constituents of a vector cannot, in general, be construed as *causal* explanations. So, as chapter 9 wanted to know, if the explanation of compositionality that S-architectures offer isn't causal, *what sort of explanation is it?*

Smolensky has said a lot about this in his papers, but I honestly don't understand a word of it. For example: "Tensor product constituents [what I'm calling "derived" constituents] play absolutely indispensable roles in the description and explanation of cognitive behavior in ICS [integrated connectionist symbolic architectures]. But these constituents do not have a *causal* [sic] role in the sense of being the objects of operations in algorithms actually at work in the system. These constituents are in this sense *acausally explanatory* [sic][11] (249). If, like me, you don't know what sort of thing the acausally explanatory explanation of "cognitive behavior" would be, I don't expect you'll find that such passages help a lot, since though they tell you what acausal explanations *aren't* (viz., they aren't causal), they don't tell you what acausal explanations *are.* There are, notice, quite a lot—say, infinitely many—of domains of "objects" that aren't ever tokened in anybody's head but which correspond bi-uniquely to the representations that mental processes work on; every scheme for Gödel numbering the mental representations provides one, as Smolensky himself observes. Do all these "objects" acausally explain cognitive behavior? If not, why not?

Digression on singing and sailing   In an enigmatic footnote to the passage I just quoted,[6] Smolensky asks rhetorically whether it is "so incomprehensible that Simon's and Garfunkel's voices each have causal consequences, despite the fact that neither are 'there' [sic] when you look *naively* [sic] at the pressure wave realizing *The Sounds of Silence?*" (284)[7] What makes this puzzling is that the passage in the text wanted to *deny* what the footnote apparently wishes to *assert*; namely, that theories that appeal to derived constituents thereby invoke causes of the events that they explain. Smolensky himself appears to be having trouble figuring out which explanations are supposed to be the "acausal" ones. I do sympathize.

In any event, the answer to the rhetorical question about Simon and Garfunkel is "yes and no." That the several voices should have their several effects is *not* incomprehensible *so long as you assume that each of the voices makes its distinct, causal contribution to determining the character of the waveform to which the ear responds.* Alas, the corresponding assumption is *not true* of the contribution of trees and their constituents to the values of the vectors that encode them in S-architectures.

Perhaps a still homelier example will help to make things clear. One normally sails from A to B along a vector that corresponds (roughly) to the sum of the force that the wind exerts on the boat and the force that the current exerts on the boat. The wind and the current make their respective causal contributions to determining your track *jointly* and *simultaneously* in the ways the resultant vector expresses. The counterfactuals fall into place accordingly: Increase the force of the wind and, ceteris paribus, the value of the vector changes in the wind's direction; increase the force of the current and, ceteris paribus, the value of the vector changes in the current's direction. And likewise, mutatis mutandis, for Simon's and Garfunkel's respective contributions to determining the form of the sound wave your hi-fi set produces and your ear responds to. This is all fine, well understood, and not at all tendentious.

But notice that this story is quite *disanalogous* to the way in which, on Smolensky's account, vector values vary with the constituent structures of the trees that they encode. For, to repeat, the wind and current determine the value of the vector that one sails along *by exerting their respective causal forces on the boat.* So, it goes without saying, there must *really be* winds and currents acting causally upon the hull, and there must be causal laws that control these interactions. Otherwise, that the boat sails along the vector that it does is an utter mystery. It's only because properties like *being a force four wind* are occasionally instantiated and locally causally active that the vectors one sails along ever have the values that they do. *If the wind and the current were just "imaginary"*—specifically, causally nugatory—boats without motors wouldn't move.

In contrast, remember, the trees that vectors encode need *never* be tokened, according to Smolensky's theory. A fortoiri, the theory does *not* countenance causal interactions that involve tree tokens, or causal laws to govern such interactions. The question thus arises for Smolensky (but not for Simon or Garfunkel or for me out sailing) is *how*—by what causal mechanism—the values of the putative vectors in the brain come to accord with the values that the encoding algorithim computes. On the one hand, Smolensky agrees that explaining compositionality (etc.) requires tree-talk. But, on the other hand, he concedes that S-achitectures acknowledge no causal transactions involving trees. Smolensky really does need an explanation of how both of these claims could be true at the same time. But he hasn't got one.

To return to the main thread: Let's grant, to keep the argument going, that there are indeed acausal explanations, and that derived constituents are sometimes invoked in giving them. Granting all that, can't we now stop the philosophy and get to the cognitive science? No, Smolensky is *still* not out of the woods. The problem resides in a truism that he appears to have overlooked : *Bi-uniqueness goes both ways.*

Apparently Smolensky is reasoning as follows: That vectors are bi-uniquely derivable from trees licenses an architecture according to which vectors causally explain some of the facts and trees acausally explain the rest. But then, by the very same reasoning, that trees are bi-uniquely derivable from vectors should license an architecture according to which trees causally explain some of the facts (compositional phenomena, as it might be) and vectors acausally explain the rest. Sauce for the goose, sauce for the gander, after all. If unique derivability in one direction suffices to transfer the explanatory power of a tree architecture to vectors, unique derivability in the other direction should correspondingly suffice to transfer the explanatory power of a vector architecture to trees; prima facie, the total of explanation achieved is the same in either direction.[9] I am not, please note, offering this observation either as an argument for tree architectures or as an argument against vector architectures; it's just a reductio of Smolensky's assumption—which is, anyhow, wildly implausible on the face of it—that explanatory power is inherited under unique derivability.

The plot so far: Smolensky thinks the best account of compositionality is that it's causally explained by vectors and acausally explained by trees. But he explicitly supposes that trees and vectors are biuniquely derivable, and apparently he implicitly supposes that explanatory power is inherited under biunique derivation. So it looks like, on Smolensky's own principles, the account according to which compositionality is causally explained by trees and acausally explained by vectors *must* be as good as the one that Smolensky actually endorses. Smolensky needs, to speak crudely, to get some asymmetry into the situation somewhere. But there doesn't seem to be any place for him to put it.

Patently, Smolensky can't afford to give up the inheritance of explanation under biunique derivability. That would invite the question of why the derivability of vectors from trees should be *any reason at all* for supposing that vectors can explain what trees do. Likewise, he can't afford to give up the biuniqueness of the algorithm that encodes trees as vectors. For, on the one hand, if you can't derive a unique vector corresponding to each tree, then clearly the vector notation doesn't preserve the structural distinctions that the tree notation expresses; this is trivially true, since if trees map onto vectors many to one, the vectors *thereby* fail to preserve the structural properties that distinguish the trees. And, on the other hand, if you can't derive from each vector the unique tree that it encodes, then the concept of a "derived constituent" of a vector isn't well defined and the whole scheme collapses.

But even though Smolensky needs explanation to be inherited under derivability and "is *derivable* from," to be symmetric, "is (*actually*) *derived* from" needn't be. So why, you might wonder, doesn't Smolensky say this: "The reason vectors explain what trees do but trees don't explain what vectors do is that vectors are actually derived from trees in the course of mental processing, but not vice versa." This would have the added virtue of making "derived constituent of" a sort of co-tokening relation: to get a vector token for mental processes to operate upon, you would need first to token the tree that the vector derives from. And if derived constituency is a co-tokening relation after all, then maybe S-architectures could provide for *causal* explanations of compositionality phenomena after all—in which case, Smolensky could just scrap the puzzling talk about "acausal explanations"—a consummation devoutely to be wished.

In fact, however, Smolensky doesn't and mustn't say any of that. For the mind actually to derive vectors from trees—for it actually to execute the algorithm that takes trees onto vectors—would require a cognitive architecture that can support operations on trees. But, by general consensus, S-architectures can't do so; S-architectures work (only) on vectors. That, after all, is where we started. It was, as we saw, *because* S-architectures don't support operations on trees that they have trouble explaining compositionality in the first place.

To put the same point slightly differently: If Smolensky were to suppose that vector tokens come from tree tokens via some algorithm that the mind actually executes in the course of cognitive processing, he would then have to face the nasty question of where the tree tokens themselves come from. Answering that question would require him to give up his connectionism since, again by general consesus, connectionist architectures don't generate trees; what they generate instead is *vector encodings* of trees.

It's important, at this point in the discussion, to keep clear on what role the tree encoding algorithm actually plays in an S-architecture.[10] Strictly speaking, it isn't part of Smolensky's theory *about the mind* at all; strictly speaking, it's a part of his theory about *theories about the mind.* In particular, it's a device for exhibiting the comparability of classical architectures and S-architectures by translating from the tree vocabulary of the former into the vector vocabulary of the latter and vice versa. But there is no cognitive architecture that postulates mental processes that operate on *both* kinds of representation (at least there is none that is party to the present dispute);[11] a fortiori, there is no mental process in which the two kinds of representations are supposed to interact; and no mind ever executes Smolensky's encoding algorithm in the course of its quotidian operations (except, perhaps, the mind of a theorist who is professionally employed in comparing classical architectures with connectionist ones).

So if everybody, Smolensky included, agrees that the encoding algorithm isn't really part of Smolensky's theory of how the mind works, why does Smolensky keep making such a fuss about it? It's because, since he admits that the explanation of compositionality should be couched in terms of trees and their constituents, Smolensky needs somehow to make the vocabulary of tree-talk accessible to vector theories. The function of the encoding algorithm in Smolensky's overall picture is to permit him to do so, and hence to allow the connectionist explanation of compositionality to parasitize the classical explanation. That's all it does; it has no other motivation.

The sum and substance is this: Smolensky's argument is, for all intents and purposes, that since there is exactly one vector that is derivable from each tree, then if the structure of a tree explains compositionality (or whatever else; the issue is completely general), so too does the structure of the corresponding vector. *Smolensky gives no grounds, other than their biunique derivability from trees, for claiming that vectors explain what trees do.* Put this way, however, the inference looks preposterous even at first blush; *explanation is not, in general, preserved under one-to-one correspondence;* not even if the correspondence happens to be computable by algorithm. Why on earth would anyone suppose that it would be?

In effect, Smolensky proposes that the classical theory should do the hard work of explaining compositionality, systematicity, etc., and then the connectionist theory will give the *same* explanation except for replacing "constituent" with "derived constituent" and "explain" with "acausally explain" throughout. Would you like to know why thinking *brown cow* requires thinking BROWN? Glad to oblige: It's because, since BROWN is a classical constituent of BROWN COW, it follows by definition that the BROWN vector is a derived constituent of the BROWN COW vector. And, by stipulation, if C is a derived constituent of C^*, then your thinking C^* *acausally explains* your thinking C. Smolensky's architecture offers no alternative to the classical story and adds nothing to it except the definition and the stipulation. This way of proceeding has, in Russell's famous phrase, all the virtues of theft over honest toil. Can't something be done about it?

What Smolensky really wants is to avail himself of tree explanations without having to acknowledge that there are trees; in effect, to co-opt the vocabulary of the classical theory but *without endorsing its ontology*. But that, alas, he is not permitted to do, unless he is prepared to recognize the resulting theory as merely heuristic. (It is, of course, perfectly alright to talk about the Sun's orbit around the Earth for purposes of doing navigation; but that's sailing, not science.)

If, for an example that I hope is untendentious, you want rock-talk in your geological theory—if you want, that is, to frame your geological generalizations and explanations in terms of rocks and their doings—you will have to admit to rocks as part of the actual causal structure of the world. What you are *not* allowed to do is borrow rock-talk when you want to explain what it is that holds Manhattan up and *also endorse a rock-free ontology when you come to saying what the world is made of*; that would be cheating. After all, how *could* rocks explain what holds Manhattan up if there aren't any rocks? It is not, notice, a respectable way of avoiding this question to reply that they do hold it up, only acausally.

Likewise, you are not allowed to borrow the idea that the constituent structure of classical mental representations is what explains the compositionality of thought and also deny that there are mental representations that have classical constituent structure; that is cheating too. How *could* classical constituents explain why thought is compositional if thoughts don't have classical constituents?

Smolensky proceeds as though the choice of an explanatory vocabulary and the choice of an ontology were orthogonal parameters at the cognitive scientist's disposal. To the contrary: What kinds of things a theorist says there are sets an upper bound on what taxonomy his explanations and generalizations are allowed to invoke. And what taxonomy his explanations and generalizations invoke sets a lower bound on what kinds of things the theorist is required to say that there are. In this fashion, science goes back and forth between how it claims the world works and what it claims the world is made of, each kind of commitment constraining the other (to uniqueness if we're lucky).

Smolensky, it appears, would like a special dispensation for connectionist cognitive science to get the goodness out of classical constituents without actually admitting that there are any. In effect, he wants, just this once, to opt out of the duality of ontology and explanation; that's what his appeals to acausal explanation are intended to allow him to do. It's the special convenience of acausal explanations that, by definition, they carry no ontological burden; just as it's the special convenience of free lunches that, by definition, there is no charge for them. That's the good news. The bad news is that there aren't any.

Acknowledgment

I'm very grateful to Zenon Pylyshyn for his comments on an earlier draft of this chapter.

Notes

1. Just to have a label, I'll sometimes call whatever things belong to this bundle "compositional phenomena." It won't matter to my line of argument exactly which phenomena these are, or whether compositionality is indeed at the center of the cluster. Though I have strong views about both of these questions, for present purposes I don't need to prejudge them.

2. For this terminology, see Fodor and Pylyshyn (1987). What distinguishes classical architecture from local connectionism is that the former recognizes *two* sorts of primitive relations between mental representations, one causal and one structural. By contrast, local connectionist architectures recognize the former but not the latter. Fodor and Pylyshyn argue, correctly, that the problems that local connectionists have with compositional phenomena trace to this fact. Apparently Smolensky agrees with them.

3. More terminology: Smolensky calls his kind of theory an "integrated connectionist/ symbolic cognitive architecture." That, however, is a mouthful that I'd rather not swallow. I'll call it a "Smolensky Architecture" or "Smolensky Theory" ("S-architecture" or "S-Theory" for short).

4. Still more terminology. (Sorry.) For present purposes, I'll use "concept" and "mental representation" pretty much interchangably; I'm supposing, in effect, that concepts are interpreted mental representations. I shall, however, want to distinguish between a concept (or mental representation) and its semantic value (e.g., the individual that it denotes or the property that it expresses). I'll write names for concepts in capitals (thus "RED" denotes the concept RED) and I'll write names for the semantic values of concepts in italics (thus, the concept RED expresses the property *red.*) These conventions tacitly assume a representational theory of mind; but that's common ground in the present discussion anyway.

    The reader should bear in mind that "RED," "BROWN COW," and the like are supposed to be *names* of concepts, not structural descriptions. The notation thus leaves open whether "BROWN COW" (or, for that matter, "RED") names a complex concept or an atomic one. It is also left open that mental representations are kinds of vectors and have no constituents.

5. This is concessive. If you think that, in the long run, even what passes as syntactic constituency must be semantically defined, so much the better for the line of argument I'm about to develop.

6. The next three paragraphs differ substantially from the original published version. Smolensky has often complained that the role vectors play in his theory is just like the role that they play in untendentious scientific explanations in (e.g.) mechanics. So why is everybody picking on Smolensky? This question deserves an answer; the discussion in the text provides it.

    By the way, the discussion here connects with what's said in passages in chapter 9: that the "normal decomposition" of a vector is, ipso facto, decomposition into factors the ontological and causal status of which the theory acknowledges. The sailor factors the vector he sails on into components for wind and tide because he believes (correctly) that both are causally implicated in determining his course.

    I am indebted to a discussion with Smolensky, Brian McLaughlin, and Bruce Tessor among others for these revisions.

7. I have made inquiries. It would appear that *The Sounds of Silence* is some popular song or other, of which Simon and Garfunkel are among the well-known expositors. I don't imagine that the details matter much.

8. Sometimes Smolensky writes as though S-architectures offer *reductions* of classical architectures (see, e.g., 1955b, 272: "the symbolic structure of [classical] representations and the recursive character of the functions computed over these representations have been *reduced to* [my emphasis] tensor product structural properties of activation vectors...." See also 1995a, passim). But that can't really be what he intends. Causality is *preserved* under reduction; it couldn't be that water reduces to H_2O *and* that H_2O puts out fires *and* that water doesn't. But that vectors are causes *and trees aren't* is Smolensky's main claim.

    Chapter 9, trying hard to square what Smolensky says about reduction and explanatory levels with his insistence that classical explanations are acausal, suggests that maybe Smolensky is really some sort of closet epiphenomenalist. I still wouldn't be in the least surprised.

9. To be sure, the two treatments would differ in the way they divide the burden between causal and acausal explanation; and in principle, that might allow one to choose between them. Perhaps, for example, some kinds of facts are intrinsically suited for causal explanations, while others are, by their nature, best acausally explained. But who knows which kinds of facts are which? In particular, what could justify Smolensky in assuming that facts about compositionality are of the latter sort?

10. I think, by the way, that Smolensky *is* clear on this; that is why he conceeds that "in the classical, but not the ICS architectures ... constituents have a causal role in processing" (236). I therefore regard this part of the discussion as untendentious and expository. The polemics start again in the next paragraph.

11. Why not a "mixed" theory, as a matter of fact? Sure, why not. But the claim that *some* mental processes are vector transformations sounds like a lot less than the pardigm shift that connectionists keep announcing. I suppose, for example, Smolensky would agree that if a cognitive architecture allows tree operations *as well as* vector operations, it should be the former that its explanation of compositional phenomena appeals to. It is, after all, exactly because he admits that trees are the natural way to explain compositionality that Smolensky feels compelled to invoke explanations in which they figure acausally.

Chapter 11

There and Back Again: A Review of Annette Karmiloff-Smith's *Beyond Modularity*

These days, hordes of people are interested in the idea that aspects of cognitive architecture may be modular. I know at least two or three (people, not hordes or aspects), and there may be others. But "modularity" means different things on different tongues. In this chapter, I want briefly to distinguish between some versions of modularity theory that are currently in play. Then, I will discuss one of them in critical detail.

There are four essential properties connected with the notion of a module: Unless you believe that at least some mental entities instantiate at least two of them, you are not in the modularity camp according to my way of choosing sides.

1. Encapsulation Information flow between modules—and between modules and whatever unmodularized systems the mind may contain—is constrained by mental architecture. "Constrained by mental architecture" means "not cognitively penetrable" (Pylyshyn, 1984): You can't change such an arrangement (just) by fooling around with someone"s beliefs and desires. In particular, architectural arrangements are (relatively) insensitive to instructional variables in experimental tasks. The persistence of illusions is the classical instance. Convincing the subject that the Muller-Lyre effect is illusory doesn't make the apparent difference between the length of the lines go away.

2. Inaccessibility In effect, the inverse of encapsulation. Just as information about beliefs and desires can't get into a module, so the information that is available to its computations is supposed to be proprietary and unable to get out. In particular, it is supposed not to be available for the subject's voluntary report.

3. Domain specificity The information and operations by which a module is constituted apply only in the module's proprietary domain. Concepts and processes may thus be available for language learning, or face recognition, or space perception which are not likewise available for balancing one's checkbook or deciding which omnibus to take to Clapham.

Table 11.1

	NC	AKS	JAF
encapsulated	don't care*	yes and no #	yes
inaccessible	yes	yes and no%	yes
domain-specific	yes	yes	yes
innate	yes	yes and no@	yes

Note: NC is Chomsky, AKS is Karmiloff-Smith, and JAF is me.

4. *Innateness*   The information and operations proprietary to a module are more or less exhaustively "genetically preprogrammed" (whatever, exactly, that means).

People who agree that some mental processes are modular may, nonetheless, differ appreciably in their views about the encapsulation, accessibility, domain specificity, and innateness of even their favorite candidates. Table 11.1 shows a rough sketch of the way three currently active theorists distribute, all of whom think of themselves as promodule in some sense or other. Comments:

*Chomsky, in some of his moods, dislikes the whole information-processing view of mental operations. If the mind isn't an information processor at all, then the question of whether it's an encapsulated information processor doesn't arise.

A proposed principle of the ontogeny of cognition: Mental processes *become* encapsulated in the course of cognitive development (perhaps through overlearning). So they are encapsulated *synchronically* but not *diachronically*. I'll refer to this putative process as *modularization* (my term, not Karmiloff-Smith's).

%Another proposed principle of cognitive development: Modularized information becomes increasingly accessible over time as a result of an "epigenetic" process of *representational redescription* (the "RR" theory).[1] I'll refer to this as a process of demodularization (again, my term).

@What's innate: Some domain-specific information and "attentional biases"; and, presumably, the psychological mechanisms that underlie the putative epigenetic processes. But neither encapsulation nor accessibility are themselves genetically preprogrammed.

An aside about attention: It's a recurrent theme in *Beyond Modularity* (BM) and also in Elman et al. (see chapter 12) that "There must be some innate component to the acquisition of [e.g.] language [but] ... this does not mean that there has to be a ready-made module. Attention biases and some innate predispositions could lead the child to focus on linguistically relevant input and, with time, to build up linguistic representations that are domain specific" (36). This emphasis on innate attentional biases is not widely shared by modularity theorists. It strikes me as unpromising, and I won't discuss it in what follows. In neither Karmiloff-Smith's book nor Elman's is it explained how one could have a disposition (innate or otherwise) to concentrate on Xs unless one *already* has the concept of an X. ("Pray, attend to the passing flubjumbs." "Can't." "Why not?" "Don't know what a flubjumb is.") Postulating innate attentional biases doesn't dispense with the postulation of innate conceptual content; it just presupposes it.

It may be that, in passages like the one just quoted, Karmiloff-Smith is only suggesting that it would be a help to the child to be (differentially) interested in speechlike *sounds*. That's quite plausible, in fact; but it doesn't even begin to explain *how* someone who is so biased manages "with time, to build up linguistic representations that are domain specific." As far as anybody knows, you need innate conceptual content to do that; indeed, as far as anybody knows, you need great gobs of it. (I am disposed to attend to the speech sounds that German speakers make; but I find learning German *very hard* for all that.)

So much for some current kinds of modularity theories. Perhaps I should say at the outset that I think you'd have to be crazy to bet much on which, if any, of them is true. The study of mental architecture is in its infancy, and it looks to be developing *very slowly*. My modest ambition in what follows is just to indicate some doubts about Karmiloff-Smith's view. And, even here, I'm not going to argue for anything so positive or decisive as that she is plain wrong about modularity. I will, however, try to show that the ways she sets out her view, and the ways she undertakes to defend it, are insensitive to certain distinctions that a cognitive architect really ought to observe. And that, when this is all cleared up, what's left may after all be true—who knows?—but, as things stand, neither the arguments for modularization (the thesis that cognitive architecture becomes increasingly modular with development) nor the arguments for demodularization (the thesis that information in modules becomes increasingly accessible with development), are persuasive.

Modularization

I won't treat modularization at length since Karmiloff-Smith makes practically no positive case for it except for remarking that the plasticity of the infant's brain militates against the thesis that its cognitive architecture is innately preformed. I think this consideration cuts little ice against nativism, since modularization, if there is such a process, might be maturational, and the course of maturation might itself be genetically specified. (Something like this is surely true for the development of secondary sexual characteristics, for example; why, then, couldn't it be true of brain structures?) And anyhow, nobody knows what the neural plasticity of the infant's brain *means*. Nobody has any idea, for example, whether the infant's brain is plastic *in respects that affect cognitive architecture*. (For more on this, see chapter 11.)

No doubt, Karmiloff-Smith is right to insist that nothing we know actually rules out progressive modularization as a process in cognitive development. As she says, though "[i]t is true that *some* [*sic*] genetically specified predispositions are likely to be involved ... such a claim should not automatically negate the *epigenetic* [*sic*] influence of the sociocultural environment on the [child's cognitive] development" (129). As far as I can see, however, there is no positive evidence that a processsess of modularization does, in fact, occur in ontogeny; and, lacking such evidence, steady-state is surely the least hypothesis.

Karmiloff-Smith does, however, make a case for demodularization; doing so occupies most of her book. I now turn to this.

Demodularization

It's essential to Karmiloff-Smith's story that there are interesting, endogenously driven reorganizations of cognitive domains that typically occur *after* a child achieves "behavioral mastery" in the domain—hence, after the point at which most developmental cognitive psychologists lose interest in cognitive development. Among these is a purported increase in the accessibility to voluntary report of information that was previously proprietary to a modularized system. Here's one of her parade examples: On the one hand, "once young children are beyond the very initial stage of language acquisition and are consistently producing both open-class and close-class words ... there can be no question that at some level these are represented internally as *words* [*sic*]" (51). But, on the other hand, "When asked to count words in a sentence, young children frequently neglect to count the closed-class items. When asked directly if 'table' is a word, they agree; but when asked if 'the' is a word, they answer in the negative" (51–52). What explains such findings, according to Karmiloff-Smith? "The RR model posits this developmental progression can be explained only by invoking, not one representation of linguisic knowledge, to which one either has or does not have access, but several re-representations of the same knowledge, allowing for increasing accessibility" (54).

This sort of case is very close to the heart of Karmiloff-Smith's attempt to get beyond modularity, so I want to look at it rather closely. I'll make two claims, which will form the substance of the discussion to follow: first, there is actually no evidence that the accessibility of modularized information increases over the course of cognitive development; and second, even if accessibility does increase, the *redescription* of the modu-larized information wouldn't explain *why* it does.

Let's start with the claim that the six-year-old child, who explicitly reports that "the" is a word, has access to intramodular representations that are inaccessible to the three year old. The three year old, we're assuming, marks the word boundary between "the" and "boy" in the course of such modularized "on-line" tasks as parsing an utterance of "the boy runs," but nonetheless denies that "the" is a word when asked for metalingustic judgments. So, plausibly, *something* is accessible to the older child that the younger one can't get at. The crucial question is whether it's the accessibility of *intramodular* representations—that is, of information *inside* the module—that has changed in the course of development.

(In passing: I say it's *plausible* that it's something about the accessibility of linguistic information that changes between ages three and six, not that it's apodictic. It could be that all that happens is that a certain linguistic confusion gets resolved: Young children think the word "word" means *open-class word*, whereas older children know better. It wouldn't be awfully surprising if three year olds are confused about this; most words *are* open-class words, after all, and it's likely that the examples that the child learns the word "word" from are themselves open-class. "Cat" and "mat," unlike "and" or "of," are *prototypical* words, and there's plenty of evidence that children generallly learn to identify relatively prototypical instances of a category before they learn to identify relatively marginal examples. I propose, however, to be concessive for the sake of the argu-ment, and just take for granted that the developmental change has *something* to do with alterations of accessibility.)

There is, nonetheless, a perfectly plausible alternative to the theory that what has become more accessible is an intramodular representation. To see this, let's consider for a moment what the putative language module does when it's running as an input parser. I assume—and I assume that Karmiloff-Smith would do so too—something like the picture in figure 11.1. The function of the language parser is to map incident utterance tokens, specified acoustically, onto the sorts of objects that linguists call "structural descriptions." Just what a structural description is depends, to a certain extent, on which linguist you talk to, on which day of the

Figure 11.1

week. But there's a broad consensus that it must consist in a mental representation of the input utterance at several *levels of linguistic description*. Standard candidates for the relevant levels include the following bare minimum: a phonetic representation, a phonological representation, a lexical representation, and a representation of a syntactic "constituent structure" tree.

As I say, this list is a bare minimum. But it will do for our present purposes, given just one further architectural assumption (which is also not in dispute in the current discussion), namely, that the parser computes the structural descriptions of utterances of sentences on the basis of its internally represented grammatical information about the language from which the sentences are drawn. For example, the parser computes that the structural description of "the boy ran" is (something like):

$$(((\# \text{THE} \#)_D \ (\# \text{BOY} \#)_N)_{NP} \ (\# \text{RAN})_{VP})_S$$

It is able to perform this computation because it internally represents such information as that "the" is an article, that "boy" is a noun, that "ran" is an intransitive verb, that sequences of the form (DN) are noun phrases, etc. Presumably, language learning consists in large part in supplying the module with this sort of information. (I leave open, for the moment, whether grammatical information in the parsing module is represented "explicitly" or "procedurely"; we'll return to this issue presently.)

So much for background. Consider now the question whether the three-year-old child who has achieved "behavioral mastery" (e.g., who is able fluently to parse utterances of the sentence "the boy ran") has off-line access to the information that "the" is a word. The point to which I want to draw your attention is that this question is ambiguous. For, according to the present picture, the information at issue is represented *twice* in the three-year old's head. On the one hand, each time the child succeeds in parsing a sentence that contains "the," he gets a structural description that says that "the" is a word in that sentence. And, on the other hand, part of the information about English, that is presumably specified by the grammar that's in the parser, is a lexicon (i.e., a list of the English words the child knows) which includes the word "the." So now, when you ask whether the child has off-line access to his internal representation of the fact that "the" is a word in "the boy ran," *do you mean the representation in the module, or do you mean the representation in the structural description?*

Or, to put the same question the other way around, when the six-year old exhibits explicit metalinguistic awareness that "the" is a word in "the boy ran," has the change occurred in his off-line access to what's in the module, or in his off-line access to what's in the structural description? This matters a lot for Karmiloff-Smith, because it's only if the right answer is the first one that she has produced a bona fide case of developmental demodularization. Suppose, on the other hand, that what happens in cognitive development is that over time children get increasing off-line access to the content of the structural descriptions that their modules compute. Then, for all the arguments we've got so far, there is no such thing as demodularization, and hence there is no developmental alteration of intramodular accessibility for a representational redescription theory to explain.

Now, I don't know for sure which, if either, of these stories *is* true. It seems clear, in fact, that neither could be the whole truth, because neither explains why the shift of accessibility of information about word boundaries should be specific to closed-class items.[2] For what it's worth, however, here are two straws in the wind to think about.

1. Off-line access to information in structural descriptions is apparently top-down even for adults. For example, S's response to syllables is faster than it is to phones in phoneme monitor tasks (Savin and Bever, 1970). It wouldn't be surprising if ontogeny recapitulates this asymmetry. If so, then you'd expect children to be able to report relatively high-level facts about sentences (e.g., facts about their grammatical structure and meaning) before they can report relatively low-level facts about sentences (e.g., facts about where the word boundaries are).

2. Information about word boundaries is represented both in structural descriptions and in the module. That's why Karmiloff-Smith's observations about children's metalinguistic access to word boundaries provides only equivocal support for demodularization. But there is some information that is represented in the parser but *not* in the structural descriptions that it outputs: for example, specifications of grammatical rules. Thus the structural description of "the window was broken by the rock" tells you that it's a passive; but it doesn't tell you by what operations this (or other) passives are constructed.[3] My point is that facts about the accessibility of such information provide a crucial test for the demodularization thesis. If what's going on in development is that module-internal information is becoming generally accessible, that should be true inter alia of module-internal information that is not represented in structural descriptions. In fact, however, the data are otherwise. Information that is plausibly in the module but not in structural descriptions *never* becomes accessible for metalinguistic report; not at six and not at sixty either. If you want to know how passivization works, you have to take a linguistics course.

Notice that this argument goes through even if it's assumed that the information that's in the module is procedural (that it's "know how" rather than "know that"). English speakers can't report a *procedure* for forming passives any more than they can report a *rule* for doing so.

The bottom line, so far, is that although it might be true that the walls of modules become relatively transparent in the course of cognitive development, there isn't actually any positive reason to believe that they do. In all the cases I can think of—and, as far as I can discover, in all the cases that Karmiloff-Smith discusses—the data don't distinguish the theory that it's information *in the modules* that becomes available for report from the theory that it's information *in the modules' output* that does. And, when there are data that do bear on this distinction, they favor the latter hypothesis over the former.

Redescription

Suppose, however, that I'm wrong about all this: Suppose that module-internal information does become increasingly accessible over time. What about the idea that it is some epigenetic process of "representational redescription" that effects this change?

This idea is Karmiloff-Smith's leitmotif. I've already quoted her remark about the putative shift in accessibility of word boundaries, that "the RR model posits this developmental progression can be explained only by invoking, not one representation of linguisic knowledge, to which one either has or does not have access, but several *re*-representations of the same knowledge, allowing for increasing accessibility." This sort of point is reiterated throughout her book. Thus, discussing the modularity, or otherwise, of the child's theories about the minds of conspecifics, she remarks that "the general process of representational redescription operates on the domain-specific representations of [the] theory of mind ... just as it does in other domains of cognition, to turn [them] into data structures available to other processes...." (136–137). And, discussing the child's implicit theory of the mechanics of middle-sized objects, she says, "My point is that coherently organized information about [physical] objects is first *used* by the infant to respond appropriately to external stimuli. [But] despite its coherence, it does not have the status of a 'theory.' To have theoretical status, knowledge must be encoded in a format usable outside normal input/output relations. It is these redescriptions that can be used for building explicit theories" (78). In short, "a crucial aspect of development is the redescription of ... knowledge into different levels of accessible, explict formats" (163). Summarizing all this, she says that "the human mind exploits its representational complexity by re-representing its implicit knowledge into explicit form" (191).

Now, I admit to finding such passages hard to construe. Here's my problem: Karmiloff-Smith appears to be claiming that it's the child's changing of his representational formats—in particular his changing from formats that are less accessible to formats that are more so—that explains the putative ontogenetic changes in the availability of intramodular information to explicit report, and generally, to operations outside the module's proprietary domain. But that can't be right as it stands; differences in format, in and of themselves, *can't* explain differences in accessibility. The point is sufficiently banal. Is the information that the cat is on the mat more, less, or equally accessible when it's expressed in English than when it's expressed in French? That's a nonsense question on the face of it; it's *more* accessible if you're a monolingual English speaker, but it's *less* accessible if you're a monolingual French speaker, and it's *equally* accessible if you are fluently bilingual. No "format" is either accessible or inaccessible as such. So no story about changing formats is, in and of itself, an explanation of changes in accessibility.

Likewise for explicitness. Every representation is explicit about *something*; pictures are explict about (e.g.) shape and color, sentences are explicit about (e.g.) tense and negativity. Correspondingly, if you change format from pictures to words, some aspects of what's represented become less explicit, and some aspects become more so. But no system of representation is explict or inexplicit *as such*; no system of representation is more—or less—explicit than any other *across the board*.

Representational formats, in short, aren't "accessible" or "explicit" per se; they're only "accessible" to something or "explicit" about some or other of the information that they represent. So, suppose that the child does change representational formats in the course of cognitive development. That wouldn't explain the (putative) fact that intramodular information becomes increasingly available in the course of cognitive development; it just raises the question of *why* the information is more accessible in the new format than it was in the old one. Indeed, it raises the question of why the (putative) change of the representational format should affect the accessibility of the representations *at all*.

I don't think that Karmiloff-Smith ever faces these sorts of questions squarely, but she does drop a number of hints; they're implicit (or, anyhow, inexplicit) in the several by no means coextensive claims that she makes about what *kinds* of redescriptions go on in the course of development. I'll discuss four of these. The moral will be that none of Karmiloff-Smith's stories about how intramodular representations are redescribed in the course of ontogeny provides a plausible account of why such representations should become increasingly accessible over time.

1 "Procedural" Representation

In the early stage of the development of a domain-specific cognitive system, the information relevant to task performance is encoded "procedurally."

Thus "at level I, representations are in the form of procedures for analyzing and responding to stimuli in the external environment.... Information embedded in level-I representations is therefore not available to other operators in the cognitive system.... A procedure *as a whole* [sic] is available as data to other operators; however *its component parts* [sic] are not" (20).

It's again not clear to me how Karmiloff-Smith wants this story to go. It sounds as though she thinks that it's the (presumed) fact that the information is procedurely represented that *explains* why its component parts are not "available to other operators in the cognitive system." But, surely, it doesn't.

Though it gets thrown around a lot in cognitive science, the notion of a procedural representation isn't itself transparent. At its least tendentious, however, a procedural representation is just a representation of a procedure. This is the construal that's suggested by examples like sentence parsing. It may be that what underlies the child's ability to assign syntactic forms to utterances is something like an algorithm for mapping sentence tokens under acoustic representation onto their structural descriptions. A grammar of the child's language may be "explicit" in the parser, or it may be merely "implicit" in the algorithm in the sense that, whereas the latter contains "declarative" representations (like " 'the' is a word"), the former contains "imperative" representations (like "if you find a phonological sequence $\#/t//h//e/\#$, label it a token of the word-type "the"). Notoriously, parsers and grammars needn't be trivially inter-convertible. Going in one direction, there are grammars for which parsers aren't constructible; and, going in the other direction, there needn't be any fact of the matter about which of an infinity of extensionally equivalent grammars a given parser realizes. So, on the construal of "procedural representation" as "representation of a procedure," there really can be something of substance to the distinction between procedural representations and others.

But that's no use to Karmiloff-Smith. For there's no obvious reason that parsing algorithms qua representations of procedures should be either more or less accessible for the subject to report than grammars qua representations of languages. Nor, as far as I can tell, is there any respectable sense in which parsing algorithms are either more or less *explicit* than grammars; they're just explicit about different things. So, for all I know, it could be true that children start out having just a parsing algorithm and end up having rules of grammar as well; and, for all I know, it could be true that children are metalinguistically explicit about the rules of their grammar but not about their parsing algorithm. But neither of these claims would explain the other, so if both are true, then Karmiloff-Smith has three worries on her hands rather than just the one she used to have.

She used to have to explain why intramodular information gets more accessible over time; now she has to explain that and also why the format of intramodular information becomes less procedural over time, and why, in general, procedural information should be supposed to be less accessible than grammatical information.

To repeat: Karmiloff-Smith seems to think that if she could explain why intramodular information becomes deproceduralized in the course of development, that would explain why it becomes increasingly accessible in the course of development. But it wouldn't. *Here as elsewhere, format as such is simply neutral with respect to accessibility as such.*

There is, to be sure, another notion of procedural representation, which I mention just to get it out of the way; it can't be what Karmiloff-Smith has in mind. On this view, a procedural representation isn't a representation of anything; it's not even the representation of a procedure. To have a procedural representation of how to solve a certain sort of problem is just: to be able to solve the problem. In this (rather strained) sense, I have access to a procedural solution of the problem of how to raise my arm; that is, I am able to raise it. Maybe my ability to ride a bike is procedural in that sense; I just *can* ride it, and no representations are invoked in explaining how I do.

But, as I say, this can't be what Karmiloff-Smith has in mind since if procedural representations aren't representations, then development can't be a process of *re*-representation. Development can only change the format of information that *has* a format; which, according to the present construal, procedural representations do not.

2 "Redescription" as Inductive Generalization
Here's a second, and I think quite different, kind of story that Karmiloff-Smith tells about how it is that "representational redescription" accounts for the demodularization of information in the course of cognitive development—one that doesn't invoke the notion of procedural representation. The child starts with representations of a variety of discrete cases; subsequently, he generalizes over these, arriving at rules that (purport to) apply to all the cases of a certain kind.

This is a kind of developmental progression that pretty likely does occur; for example, it's presumably what generates the familiar phenomenon of "U-shaped" developmental curves. Here's a case that's well known from the literature on the child's developing mastery of past-tense morphology. Ontogeny often exhibits the following phases:[4]

    phase 1: go/went
    phase 2: go/goed
    phase 3: go/went

Apparently the child at phase 2 is operating on a generalization about the past-tense which the child at phase 1 hasn't formulated: roughly, that you make a past tense verb by sticking "-ed" on the end. In fact, at phase 2 the child takes this generalization to be more reliable than it actually is; he applies it indiscriminately to strong and weak verbs. By phase 3, he has learned that "go" is an exception to the "add -ed" rule of past-tense formation.

But though this sort of thing no doubt happens, it too is no help to Karmiloff-Smith.

(i) Technical quibble   Inductive generalization isn't really the "redescription" of old information; it's the addition of new information. Universal generalizations are stronger than the singular statements that instance them. For one thing, they govern new cases. (If you know the "add -ed" generalization, then you predict that "fricked" is the past tense of "to frick." If you don't know the generalization, you can't make this prediction.)

(ii) More important, there's every reason to think that this sort of "redescription" is entirely intramodular.   Adult speakers, by assumption, internally represent the generalizations about their lanaguage of which children represent only the instances. But adults can't articulate such generalizations any more than children can. (Thus most adults speakers know explicitly that "walked" is the past tense of "walk" and that "vended" is the past tense of "vend"; but they can't tell you why it is that in one case the past-tense morpheme is sounded /t/ and in the other case it's sounded /id/.) There is, in short, no reason to suppose that the kind of "redescription" that goes on in formulating internal representations of such inductive generalizations is a move in the direction of increased accessibility. A fortiori, there's no reason to suppose that it's a move in the direction of demodularization.

(iii) Redescription is like theory construction; the child is a "little theorist"   "Normally developing children not only become efficient users of language; they also spontaneously become little grammarians" (32); "[they] go beyond the input/output relations and reflect metalinguistically on the word" (33). Likewise, "young children spontaneously come to theorize about the physical world ... by the internal process of representational redescription which abstracts knowledge the child has already gained from interacting with the environment" (78). And so forth in many places in the book.

I'll be brief about this proposal. I have no doubt that (most, many, for all I know, all) children reflect upon and theorize about the world; and I have no doubt that their doing so comes as naturally to them as flight to butterflies. Nor do I doubt what Karmiloff-Smith is often at pains to insist on, that what children do when they theorize is in many respects strikingly similar to what grown-up scientists do when they are professionally engaged; that, for example, there's a methodological premium on simplicity, generality, and conservatism in both cases.

But there is, to my knowledge, *no evidence at all* for the claim that, in children, theory construction is a process of demodualization. In particular, there's no evidence at all that the data that prompt the child to theorizing are ever intramodular.

This connects with a point I made some way back. The natural assumption is that what children theorize about is not what's represented *in* their modules, but rather what's represented in the *outputs* that their modules compute. According to this view, the child does have increasing reflective access to something that she mentally represents; and, as her access to what she mentally represents increases, so does her inclination to theorize on the content of the representations. But what the child has increasing access to is information in the structural descriptions that the modules deliver, not intramodular information per se. Otherwise grownups could introspect the grammars of their language and linguists would be out of work. That would be a Very Bad Thing; my wife is a linguist, and I am counting on her to support me in my retirement.

(iv) Connectionism I should add just a note about Karmiloff-Smith's suggestion, in the last chapter of her book, that one might explicate the notion of procedural ("I-level") representation by analogy to what goes on in neural nets. As she reads the connectionist literature, a network representation of the fact that both "boy" and "girl" are English nouns consists in the fact that there is a dimension of "the activation space [on which] 'boy' and 'girl' in all their grammatical roles line up with all the other words that we call nouns" (186). But this sort of representation of their nounhood is (merely) implicit in the sense that "it is we, as external theorists, who ... label the trajectories through weight space as nouns, verbs, subjects ... and so on. The network itself never goes beyond the formation of the equivalent of stable level-I representations.... The notion of nounhood always remains implicit in the network's system dynamics. The child's *initial* [sic] learning is like this, too. But children go on to spontaneously redescribe their knowledge. The pervasive process of representational redescription gives rise to the manipulability and flexibility of the human representations" (186).

Now, I don't know how things work in children, but this does strike me as a massive misdescription of how things work in networks (a topic in respect of which, to be sure, massive misdescription has tended to be the norm). In fact, the difference between classical and connectionist representations is orthogonal to the issue of explicitness. Here, for example, is the pertinent fragment of a network that is fully *explicit* about "boy" and "girl" both being nouns:

NOUN

boy   girl

Notice that this network is exactly as explicit about "boy" and "girl" being nouns as is the "classical" grammar of which the pertinent fragment is:

Noun → boy
    → girl

In respect to explicitness, there is *simply nothing to choose* between the two kinds or representation; the idea that classical architectures are somehow ipso facto "explicit" or that connectionist architectures are somehow ipso facto not explicit is simply confused.

I suspect that what causes the confusion is that the particular connectionist model that Karmiloff-Smith has in mind takes for granted that lexical categories like NOUN, ADJECTIVE, VERB, and the like are defined distributionally. On that view, for a word to be a noun *just is* for it to belong to a cluster of items with appropriately similar distributional properties (that's what the stuff in the quotation about "similar trajectories through state space" means). Correspondingly, a network whose connectivity mirrors the relevant distributional patterns in its input can treat NOUN and the like as "implicit" concepts in the sense that it treats them as *defined* concepts. There's nothing more to being an (English) noun than having much the same distributional properties (much the same trajectory through state space) as paradigms like "boy" or "apple."

But the question whether NOUN is in this sense an implicit concept is simply orthogonal to the question of whether cognitive architecture is connectionist. In particular, there is nothing about connectionist architecture per se that requires claiming that lexical classes have distributional definitions, and there is nothing about the classical architecture per se that requires denying that they do. That syntactic classes are distributionally defined was, indeed, a main tenet of "taxonomic" linguistics; which was classical from head to toe in its architectural assumptions.[5]

I do want to be entirely clear about the polemical situation, if I can: I am *not* saying that connectionist networks and classical grammars are other than importantly different in their treatment of language. To the contrary; there are crucial grammatical facts (particularly, facts about syntactic constituency) that classical grammars capture naturally but that network grammars can't represent *at all*. (See chapters 8 and 9.) I do claim, however, that with respect to such facts as both kinds of models *can* represent, there is no intrinsic difference in the *explicitness* with which they represent them. In consequence, even if there were some ontogenetic process by which children start out with connectionist networks in their heads and end up with classical grammars in their heads, *that wouldn't explain why there is a corresponding shift from less to more explicit metalinguistic access.*

To repeat the general moral of which the present case is an instance; differences in format don't, in and of themselves, explain differences in accessibility. So, even if there is an ontogenetic process of demodularization (which I doubt), and even if there is an ontogenetic process of representational redescription (which I also doubt), the redescription wouldn't explain the demodularization.

Conclusion

Where does all this leave us? There are two issues that everybody is trying to get clear on: What is the cognitive architecture of the adult mind? and, whatever it is, how did it get to be that way? I think the empirical results over the last couple of decades make a not implausible case that modular architecture is an appropriate idealization for some aspects of the organization of mature cognition. About the second question, however, nothing is really known; we're all just playing our hunches. Perhaps the deepest issue that divides people in the theory of cognitive development is *whether there are ontogenetic processes that affect cognitive architecture.* Karmiloff-Smith and Piaget are betting that there probably are; Chomsky and I are betting that there probably aren't. All I can tell you for sure is this: There may in fact be architectural changes in the course of the ontogeny of cognition, but nobody has found a clear case of one so far— and since the developmental plasticity of the mind has been a main theme of Anglo-American psychological speculation for a couple of centuries, I do think that fact is striking.

Notes

1. I'm not as sure as I'd like to be about what an epigenetic process is. Karmiloff-Smith provides the following gloss on Piaget's usage: "For Piaget both gene expression and cognitive development are emergent products of a self-organizing system that is directly affected by its interaction with the environment" (9). But a *self-organizing* system that is, nevertheless, "directly affected by its interaction with the environment" is, to my taste, rather like a circle that nevertheless has corners. Probably "epigenetic" just means "not primarily input driven." I shall assume that is what it means.
2. Even adults are more likely to overlook closed-class words than open-class words when (e.g.) they are hunting for typos in a proofreading task; cf. the relative inaccessibility of the

repeat in "John fed the the cat" vs. "John fed fed the cat." This is bad news for Karmiloff-Smith. Since it's common ground that adults do parse the sentences they read, and that word boundaries are marked in the structural descriptions that parsing assigns, the adult's failure to spot closed-class typos must have something to do with his access to information about closed-class items that is *in the structural descriptions*. Parity of argument suggests the same explanation for the child's lack of metalinguistic access to information about closed-class items.

3. The structural description of a passive doesn't, for example, tell you whether passivization is a lexical transformation. Indeed, whether passive is a lexical transformation is just the sort of question that linguists argue about. Clearly, linguists can't look inside their modules either.

4. Earlier claims about just how prevalent this phenomenon is have recently come in for reappraisal (see, e.g., Marcus et al., 1992). But the example will serve for expository purposes.

5. The issue about whether there can be a connectionist account of language acquisition is tacitly confounded with the issue of whether linguistic categories have distributional definitions not only in BM, but also in Elman et al., 1997. More on this in the next chapter.

Chapter 12

Review of Jeff Elman et al., *Rethinking Innateness*

Connectionism offers psychology a new kind of computational theory of the cognitive mind. Unlike the familiar "classical" models of mental processes, which borrow their architecture from digital computers, connectionist networks run in parallel; they don't honor the distinction between executive, memory, and program; and, although they are arrangements of causally interacting representational states, they don't distinguish primitive from complex representations and they don't operate by transforming symbols. These differences are substantial, so there's a serious argument going on. Readers who want an introduction to the connectionist side could do worse than chapter 2 of *Rethinking Innateness* (hereafter RI); it provides a clear, capsule tutorial. (But acquaintance with the classical approach to cognitive science is largely assumed, so RI isn't for beginners.)

There is also an argument going on about what exactly it is that classical and connectionist theories are arguing about. Nativism, associationism, empiricism, rationalism, reductionism, genetics, computer science, neuroscience, linguistics, ethology, and the mind/body problem (and I may have forgotten a few) have all somehow involved themselves in what started as just a psychologist's choice between theories of cognitive architecture. This second argument is more fun than the first; it's what RI is mostly concerned with; and it's what I focus on in this review. I'll start with what RI takes (wrongly in my view) to be the core questions. Then I'll say where I think the real issue lies. Innateness, as it turns out, is a bit of a sideshow, but we'll get to it in due course.

Brainlikeness

Advertising for connectionist models of mind invariably emphasizes their neurological plausibility. This line is developed in two ways in RI: On the one hand, there's stress on analogies between the structure of computational networks and the structure of brain tissue. In this respect, the comparisons with classical cognitive theories are often invidious, not to say sanctimonious ("[M]ost connectionist researchers are really committed to ultimate neural plausibility, which is more than you can say for most other approaches" [50]). On the other hand, the neurological data are said to be that the human cortex is, ontogenetically, an "organ of plasticity." That is supposed to be bad news for the view that innate "representational content" contributes to human cognitive development.

From the first of these claims, though it plays wonderfully in grant proposals, it might be well to avert the gaze. Connectionist models are networks of interconnected, unstructured objects called "nodes". Cortical tissues are (among other things) networks of interconnected, but not unstructured, neuronal cells. The nodes sort of look like neurons. From a distance. And you can increase the resemblance by painting them grey. Accordingly, the nodes "are likened to simple artificial neurons" (50).

At this point, we could have a serious discussion of how good, or bad, the analogy between nodes and neurons actually is. (The New York subway system is also a network of connected nodes, and it too runs in parallel.) But that would be beside the point since it turns out, as one reads further, that "the nodes in the models [are] equivalent not to single neurons but to larger populations of cells. The nodes in these models are *functional* units rather than *anatomical* units" (91). To what "larger populations of cells" are the connectionist's nodes equivalent? Deafening silence. There are exactly no proposals about how, in general, the putative functional structures and the putative anatomical structures might relate to one another. It's left wide open, for example, that nodes whose functions are quite different might correspond to anatomically similar neural structures. Or that anatomical structures that are quite different might correspond to functionally similar nodes. Or, indeed, that nodes might correspond to brain stuff that doesn't count as a unit by any independent criterion, anatomical or otherwise. All we're told is that the nodes are "distributed" over the brain stuff. Somehow.

In fact it's entirely unclear how (or whether) connectionist cognitive models are realized in actual brains, just as it is entirely unclear how (or whether) classical cognitive models are realized in actual brains. Like many other connectionist texts, RI changes its mind about this practically from page to page. By the end, it comes down to the pious "hope ... that such models will embody abstract and realistic principles of learning and change that make contact with cognitive issues, while preserving some degree of neural plausibility" (366). No doubt, "some degree of neural plausibility" would be an excellent thing for a psychology to preserve; I'll be sure to let you know if any of it should turn up.

The relation between evidence for cortical neuroplasticity and claims about innateness is more interesting and more closely germane to the special interests of RI. The authors assume, literally without argument, that "[i]n the brain the most likely neural implementation for ... innate knowledge would have to be in the form of fine-grained patterns of synaptical connectivity at the cortical level" (25). There is, however, a lot of data suggesting that much of adult cortical structure emerges from complex interactions between temporal, spatial, and neurochemical constraints and is thus, presumably, not directly genetically encoded. There is plenty of room to argue about the details; but, on balance, that's the current consensus in brain science.

The appropriate inference is therefore that *either* there isn't much genetically prespecified information represented in the cortex, or that, if there is, it isn't represented at the level of individual neurons and their synaptic connections. The authors prefer (indeed, leap to) the former conclusion; but it's hard to see why. There are, after all, levels of organization of the nervous system both larger and smaller than the interconnections between individual neurons; and any of these other levels may perfectly well be where innate—or, for that matter, *learned*—information is encoded. That one brick is a lot like another is not a reason to doubt that the structure of brick buildings is, by and large, preplanned. All it shows is that, if planning did go on, you have to look at *aggregates* of bricks to see the effects.

The way things are set up in RI, an unwary reader might suppose it's pretty clear, in general, how knowledge is encoded in the nervous system; so if we don't find the right neural connectivity when we look at the infant's brain, that shows that there's no innate content there. Nor, however, is there any learned content in the adult's brain if we insist on this line of argument. For there isn't one, *not one*, instance where it's known what pattern of neural connectivity realizes a certain cognitive content, innate *or* learned, in either the infant's nervous system or the adult's. To be sure, our brains must somehow register the contents of our mental states. The trouble is: Nobody knows how—by what neurological means—they do so. Nobody can look at the patterns of connectivity (or at anything else) in a brain and figure out whether it belongs to somebody who knows algebra, or who speaks English, or who believes that Washington was the Father of his Country. By the same token, nobody can look at an infant's brain and tell from the neurological evidence whether it holds any or all of these beliefs innately. The sum and substance is that we would all like our cognitive models to be brainlike because we all believe, more or less, that one thinks with one's brain. But wishing won't make them so, and you can't make them so by proclamation either. Nobody knows how to make a brainlike cognitive model, classical or connectionist, because *nobody knows how the cognitive brain works*.

Interactions
"We argue throughout that an exclusive focus on either endogenous *or* exogenous constraints fails to capture the rich and complex interactions which occur when one considers organisms rooted in their environment, with certain predispositions, and whose interactions give rise to emergent forms" (110); "our major thesis is that while biological forces play a crucial role in determining our behavior, those forces cannot be understood in isolation from the environment in which they are played out" (320). But this can't *really* be RI's major thesis; for who would wish to argue against it?

For example: Linguistics is the locus classicus of recent nativist theorizing. But linguists might reasonably claim to be the only cognitive scientists who have ever taken the interactionist program completely seriously. Nobody doubts—linguists doubt least of all—that what language a child learns depends on what language he hears; or that languages differ in lots of ways; or that languages don't differ in *arbitrary* ways; or that any normal child can learn any language of which he hears an appropriate sample. The goal of linguistic inquiry, as classically conceived, is therefore to provide a taxonomy of the range of variation between possible languages, and a precise account of what predispositions have to be available to a creature that is able to learn any language within this range. What more could an interactionist want in a research program?

The claim that behavior is shaped by the interaction of endogenous and exogenous information isn't worth defending because it isn't in dispute. Nobody has ever doubted it since (and including) Descartes. What's really going on in RI is a certain view of *what kinds* of endogenous information innateness can contribute to this interaction. "It is important to distinguish the mechanisms of innateness from the content of innateness.... We suggest that for higher-level cognitive behavior, most domain-specific outcomes are probably achieved by domain-independent means" (359); "We discussed innateness in terms of mechanisms and content.... We argue that representational [viz., content] nativism is rarely, if ever, a tenable position." (360–361).

Now we're getting closer.

Representational Nativism
Suppose it turned out (as, indeed, some of the data suggest it may) that human infants act for all the world as though they expect unsupported objects to fall. And suppose that the evidence was convincing that this expectation couldn't have been learned. A natural hypothesis might then be that the belief that unsupported objects fall is innate. RI wouldn't like that at all, of course; but it's important to understand that it's not the

innatetness per se that RI disapproves of. Officially all sorts of things can be innate, for all that RI cares, so long as none of them is a belief; so long as none of them has content.

Accordingly, much of RI is devoted to examples, sometimes ingenious, sometimes even credible, where what look like the effects of innate *content* can be explained as really the effects of innate *mechanisms*, typically in interaction with one another and with environmental variables. Here is a typical example: "why [do] beehive cells have hexagonal shapes.... [T]he shape of the cell is the inevitable result of a packing problem: How to pack the most honey into the hive space, minimizing the use of wax. Information specifying the shape of the cell does not lie in the bee's head, in any conventional sense; nor does it lie solely in the laws of physics (and geometry). Rather, the shape of the cell arises out of the interaction of these two factors" (320). Such interactions may occur at many levels of scale, and local constraints don't need to do all the work. "The outcomes ... typically involve complex interactions, including molecular and cellular interactions at the lowest level, interactions between brain systems at the intermediate level, and interactions between the whole individual and the environment at the highest level" (320). Any of these interactions may be biased by the genes, and to practically any degree. In short, RI will tolerate as much innate control of ontogeny as you please—but *no innate content*.

That's the official story. Actually, I doubt that mechanism vs. content is the distinction that the authors really want. Imagine, for example, an innate mechanism which, on activation (say by maturation or by some environmental trigger), causes the child to believe that unsupported objects fall. Presumably that would be just as objectionable as the innateness of the belief itself. RI is pretty unclear about all this, but I suppose that what it really wants to rule out is anything innate except general learning mechanisms; specifically, general associative learning mechanisms.

But, anyhow, what's the objection to innate content? If representational innateness is often the *obvious* theory of a creature's mental capacity, why not suppose, at least some of the time, that that's because it's the *right* explanation of the creature's mental capacity? It's not as though, in general, there's a plausible alternative on offer; often there is none. Here, for example, is what RI has to say about linguistic universals. "[Sometimes] solutions are contained in the structure of the problem space.... The grammars of natural language may be thought of as solutions of a reduction problem, in which multidimensional meanings must be mapped onto a linear (one-dimensional) output channel" (386). But no examples are given of how, even in sketch, an attested linguistic universal might be explained this way; nor is anything said about why all sign languages, which are spatially as well as temporally organized, obey much the same universal constraints that spoken languages do. And God only knows what it means to say (or to deny) that meanings are "multidimensional."

In fact, as things now stand, there isn't any explanation of why there are linguistic universals except the one that says they express innate representational contents. Moreover, there are, in consequence of several decades of investigation, detailed, abstract, and very powerful theories about which ones the universal features of languages are, and how innate knowledge of these features might interact with environmental data to produce the adult speaker/hearer's behavioral phenotype. To be sure, as RI keeps reminding us, these theories aren't "necessarily" true. But they do make the handwaving about "reduction problems" look pretty feeble.

You might suppose, so vehement is RI's rejection of content innateness, that there is something in the architecture of connectionist models that rules it out. Not at all. Assuming (what's far from obvious; but that's another story) that connectionist networks can represent cognitive content at all, they can perfectly well represent innate cognitive content inter alia. The authors themselves acknowledge this. So, the connectionism in RI on the one hand, and RI's rejection of content nativism on the other, are quite independent. If somebody proved that content nativism is false, that wouldn't tend in the least to show that connectionism is true; and if somebody proved that content nativism is true, that wouldn't tend in the least to show that connectionism is false. So, once again, what does RI have against innate representational content? This is, I think, a real aporia. You don't understand RI until you understand what's going on here.

Empiricism

"We are not empiricists," the authors of RI roundly claim (357). But all this means is what we've already seen: RI tolerates innate mechanism but jibs at innate content. By that criterion, the original empiricists weren't empiricists either. Hume, for example, thought that the mechanisms of association and abstraction are innate, and that the sensorium is too. All he objected to was innate *Ideas*. In fact, like Hume, RI is into empiricism up to its ears; it's the empiricism—not the neuroscience and not even the connectionism—that calls all the shots. RI is less a connectionist perspective on development than an empiricist perspective on connectionism, its subtitle to the contrary notwithstanding.

There are, historically, two main ways people have thought about the mind's commerce with experience. According to the picture that rationalists endorse (me too), cognitive processes are aimed at *explaining* experience. Typically, this requires solving problems of the form "What must the world be like, given that experience is such and so?" Such problems engage the mind at every level, from the local to the cosmic. What must the world be like such that these two visual points are seen reliably to move together? Answer: The points are on the surface of the same object. What must the world be like such that the red shift of a star's light is observed to correlate reliably with the distance to the star? Answer: The universe is expanding. And likewise for problems of intermediate scale: What must English be like such that people put the "John" before the "runs" when they use it to say that John runs? Answer: in English, subjects precede their verbs.

So, cognition exists to theorize about experience. One uses the concepts at one's disposal to construct the theories; that's what concepts are for. But where does one get the concepts? Some, no doubt, are bootstrapped from the results of previous theorizing, but that process has to stop somewhere (more precisely, it has to *start* somewhere). Hence the traditional connection between rationalism and content nativism. Experience can't explain itself; eventually, some of the concepts that explaining experience requires have to come from outside it. Eventually, some of them have to be built in.

The empiricist picture is much different. According to empiricists, the fundamental mental process is not theory construction but abstraction. There is a certain amount of regularity in experience, and there is a certain amount of noise. Statistical processes filter out mental representations of the noise, and what's left are mental representations of the regularities. In RI, as in much other empiricist theorizing, these statistical processes are implemented by mechanisms for forming associative connections between mental representations. Input nodes in connectionist networks respond selectively to sensory properties of impinging stimuli; other nodes form associative connections which, in effect, model correlations between the responses of the input nodes. The picture, to repeat, is not significantly different from the associative empiricism of Hume, though RI is rather more explicit than Hume was about the nonlinearity of laws of association.

To have been anticipated by Hume is not a thing to be ashamed of, to be sure. But one really should be aware of the enormous reductive burden implicit in the empiricist program, and of how unlikely it is that the required reductions could in fact be carried out. Arguably, you can form the concept of red by abstracting it out of your experience because, arguably, "red" is the name of a kind of experience. But you can't form the concept of a chair that way because a chair isn't a kind of experience, or a logical construct out of experience, or a statistical function over experience. A chair is something mind independent, which, if it's in the vicinity (and if the lights are on and one's eyes are open), causes one's experience to pattern in the familiar way. Likewise, give or take a little, for nouns and constellations. Empiricists were forever trying to reduce the concept of a cause of experience to the concept of the experiences that it causes. No wonder empiricism didn't work.

Whether mental content reduces to experiential content is what RI is really about. Sometimes, but only very rarely, RI notices this. For example, in a study that has been influential in connectionist circles, Elman "trained" a network to predict the next word in a sentence when given the preceding fragment. In the course of discussing this network, RI remarks that: "because both grammatical category and meaning are *highly correlated* [my emphasis] with distributional properties, the network ended up learning internal representations which *reflected* [my emphasis] the lexical category of the structure of the words" (343). However, the problem of language acquisition is that of how a child learns grammatical structure, not how he learns *correlates* of grammatical structure. RI does remember this distinction for almost three pages. But then it quite forgets, and Elman's network is described as having learned "distinctions between grammatical categories; conditions under which number agreement obtained; differences between verb argument structure; and how to represent embedded information" (345). These are not distributional correlates of grammatical structures, notice; they're the very thing itself. The equivocation is striking; what explains it? I suppose RI must be assuming, tacitly, that grammatical structures really *are* just distributional patterns: To be a noun is just to be (or to be distributionally similar to) a kind of word that appears to the left of a word like "runs" (or to the left of words that are distributionally similar to "runs"); and so on.

The trouble with that sort of reductive proposal, however, is that the reductions aren't ever forthcoming. In linguistics and elsewhere, it invariably turns out that there's more in the content of our concepts than there is in the experiences that prompt us to form them. (For example: There are, I suppose, nouns in Russian. If so, then *being a noun* couldn't be a property that words have in virtue of their distribution *in English*.) That there is generally more in the content of a concept than there is in the experiences that prompt us to form it is the burden of the traditional rationalist critique of empiricism. Friends more scholarly than I tell me that Leibniz had that line of criticism pretty much worked out; Kant certainly did. It has never been seriously rebutted, though empiricists often carry on as though it hadn't happened. As does RI, egregiously.

The empiricist program was in place for several hundred years; not just in psychology but also—indeed especially—in epistemology and the philosophy of science, where a lot of clever people wasted a lot of time trying to reduce "theoretical" concepts (like TABLE) to "observation" concepts (like RED and SQUARE). There were, in the course of all those centuries, *no successes at all*; literally *none*; which is to say, *not one*. For the sorts of reasons I mentioned above, pretty nearly everybody came to conclude that the failures were principled. So it looked, until just recently, as though the argument between empiricism and rationalism had finally been put to rest. The connectionists have revived it; but apparently without quite realizing that that's what they're doing, and without, as far as I can tell, having anything to add that changes the picture. So now I guess we'll have to play out the argument between empiricism and rationalism all over again. What Santayana said about history is also true of Philosophy 101: Those who haven't understood it are destined to repeat it.

Chapter 13

Review of Steven Mithen's *The Prehistory of the Mind*

What's your favorite metaphor for minds? If you're an empiricist, or an associationist, or a connectionist, you probably favor webs, networks, switchboards, or the sort of urban grid where the streets are equidistant and meet at right angles: New York's Midtown, rather than its Greenwich Village. Such images suggest a kind of mind every part of which is a lot like every other. Take a sample here and you find the concept DOG associated with the concepts CAT and ANIMAL; take a sample there and you find the concepts ANTHONY and CLEOPATRA in close connection. Though mental *content* changes as you go from place to place in the web, the *structure* is everywhere the same: It's concepts and connections wherever you look.

But there is also an older, rationalist tradition of theorizing about the mind, one that runs from Plato through Gall, Kant and the faculty psychologists, to Freud and Chomsky. It too has its proprietary metaphors, which are frequently architectural. The mind is like a building (Steven Mithen thinks it's like a cathedral). Entrance and egress are variously constrained, and so too are the paths through the interior. There are public places and private places, and places where the children aren't allowed to go; and there are places in the attic that hardly anyone but Granny ever visits. With that sort of arrangement, you might expect that information would be quite unevenly diffused among the occupants; downstairs knows things that upstairs doesn't, and upstairs doesn't know that downstairs knows them. Also, buildings are often divided into lots of little spaces, with more or less specialized functions. These differences of function generally imply differences of structure, so you wouldn't expect to learn much about how the kitchen is organized by examining the potting shed. The long and short of the building metaphor is that the whole mind doesn't always function as a system; and that when it does so, it's through the interaction of its heterogeneous parts.

The empiricist idea of a homogeneous mind has dominated psychological and philosophical thinking in the English speaking countries since the eighteenth century. But now the fashions seem to be changing, and the "modular" organization of cognition is widely promoted. Just what this amounts to is far from clear, and some of the consensus is doubtless just in terminology. Still, people in a lot of fields of research have recently come to think that some of our mental competences are isolated and specialized in ways that the web/grid/network story doesn't easily accommodate. There is even considerable agreement on which cognitive capacities are the best candidates for the status of modules. The learning and use of language is on everybody's list; Chomsky put it there, thus initiating the new fashion in cognitive architecture. Commonsense biology, commonsense physics, commonsense psychology, and aspects of visual form perception are other favorite candidates. The bidding is still open, but these five are the paradigms. If none of them is a mental module, then probably nothing is.

Isolation and specialization are the defining properties of a module, so a lot of the current discussion is about what precisely these consist in. A word about this to set the stage for Mithen's book.

First, specialization. Modules are supposed to be "domain-specific" cognitive mechanisms. To say, for example, that there is a commonsense psychology module is to claim that each of us has concepts and processes available for inferring from the behavior of conspecifics to their states of mind that are not likewise available for balancing the family checkbook or deciding which omnibus to take to Clapham. Other people are a very special part of the world; maybe one uses very special kinds of cognition when one tries to understand them.

For instance, Brentano remarked on how odd it is that one can want (need, seek, believe in) a gold mountain even though there aren't any (whereas try climbing a gold mountain if there aren't any). Apparently, the commonsense psychology module countenances relations between minds and things-that-don't-exist. It may be that only phenomena in the domain of the commonsense psychology module exhibit this logical peculiarity; philosophers think so who hold that "intentionality" is the mark of the mental. In similar spirit, Chomsky thinks that the grammatical structures of all human languages are of the same formal type, and that no communication system which failed to exemplify the universal format could engage the operation of the language module. Are there likewise eccentric constraints on the kinds of geometry that the commonsense physics module can visualize? Does physical space have to strike us as Euclidean even if it's not? Kant thought so.

What about the isolation of modules? Two kinds of ideas are current in the cognitive science literature. First, modules are said to be "informationally encapsulated"; there are supposed to be architectural constraints on the ways that information can flow between modules (and between them and whatever unmodularized cognitive capacities the mind may have; see below). The standard illustration is the persistence of perceptual illusions even when one knows that they're illusory; you know perfectly well that the moon is no bigger when it's on the horizon, but try convincing your visual perception module that it doesn't look as though it is.

Second, modules are supposed to have proprietary developmental careers. Language learning, for example, runs off largely independent of the development of visual perception. Blind children are not, in general, linguistically impaired; not even in their talk about space. Indeed, it's part and parcel of the ontogenetic autonomy of the modules that their normal development appears to make quite minimal demands on the character of a child's environment. Children don't have to be taught their native language; they just pick it up. Nor, apparently, do they have to be taught to attribute people's behavior to the mental states they're in. Children tend, quite naturally, to anthropomorphize whatever moves. What they have to learn is which things don't have minds, not which things do.

Such considerations lead to the most spectacularly controversial thesis that modularity theorists endorse: Modules are innate cognitive structures. They are, in Chomsky's phrase, "mental organs," and the autonomy of their maturation is no more surprising than the maturational autonomy of, as it might be, legs from digestive systems. You wouldn't expect fine-grained developmental measures to apply indifferently to knees and stomachs, nor would you expect the development of either to depend on special tuition. Rather, their maturation is internally driven and largely mutually independent because it is genetically preprogrammed. So too with mental modules; or so the story goes.

You might think that cognitive psychologists would reject this theory out of hand for fear of becoming unemployed. If "cognitive development" is just a cover term for the independent ontogenies of a heterogeneous bundle of mental organs, what general laws of cognitive development could there be for psychologists to discover? But this reaction is precipitous. It is, after all, not plausible that the mind could be made *only* of modules; one does sometimes manage to balance one's checkbook, and there can't be an innate, specialized intelligence for doing that. Maybe cognitive development is the story of how the child somehow gets from isolated, genetically preprogrammed, domain-specific cognitive modules to the relatively unspecialized kind of intelligence that doing one's checkbook requires.

So, finally, to Mithen, who has been reading a lot of psychologists who think that the child's cognitive development proceeds from a modularized mind to one that's "cognitively fluid" (Mithen's term). It's Mithen's daring and original suggestion that maybe this sort of ontogeny recapitulates human phylogeny. Maybe the evolution from premodern man, to early modern man, to us is the story of the emergence of cognitive fluidity from (or, rather, alongside of) the modularized components of our minds. "The differences ... are analogous to those between Romanesque and the succeeding Gothic cathedrals.... In Gothic architecture sound and light can flow freely ... unimpeded by the thick walls and low vaults one finds in Romanesque architecture.... Similarly, in the phase 3 mental architecture, thoughts and knowledge generated by specialized intelligences can now flow freely around the mind...." (71).

Well, it's a fine idea, and a fine book. At a minimum, Mithen's easy style is accessible to the casual reader who is looking for an overview, but his book is also chock full of useful archaeological and anthropological detail. A word to professionals: Do not skip the footnotes. They contain stuff that will fascinate and edify the working cognitive scientist. Mithen seduced me from the Anthony Powell that I was reading; I can't imagine warmer commendation.

But is Mithen's theory of cognitive phylogeny true? For that matter, how well does it actually fit the ontogenetic theory that he uses as a model? And, for another matter, is the ontogenetic theory that he uses as a model itself true? Mithen is three levels deep in speculation: he needs there to be mental modules; he needs the ontogeny of cognition to consist in the maturation of these modules and, eventually, in the emergence of cognitive fluidity; and he needs a phylogeny of cognition that recapitulates this ontogeny. It's a lot to ask for, and I'm not convinced that it all works.

By his own estimate, Mithen has two major epochs in the prehistory of our minds to account for. There's the transition, ten million years or so ago, from "the ancestral ape" (the presumed last common ancestor of us and chimpanzees) to "early man"; and there's the transition, one-hundred thousand years or so ago, from early man to "modern man" (in effect, to us). This taxonomy is itself not untendentious. But Mithen finds evidence, from a variety of archeological and comparative sources, that each of these major transitions implied a comprehensive increase of cognitive capacity; so comprehensive as to suggest a corresponding reorganization of the underlying cognitive architectures.

What sort of reorganization? Here's where Mithen wants phylogeny to borrow from ontogeny. The ancestral ape had a mind that ran on "general intelligence" together with, at most, a rudimentary modularized "social intelligence"; whereas, in the next stage, the mind of early man had modularized "technical intelligence" and "natural history intelligence" as well. Finally, the emergence of cognitive fluidity from a relatively compartmentalized, modular mind is what Mithen thinks distinguishes early man from us. It's this latter difference that Mithen likens to the change from Romanesque to Gothic in the passage I quoted above.

I do admire the attempt to find places for so many pieces. But it takes some forcing to make them fit, and what Mithen ends up with shows the signs of strain. For example: If cognitive ontogeny is indeed the development of a fluid mind from a modular mind, then it offers a clear precedent for the hypothesized shift from "early" to "modern" intelligence, much as Mithen says. But what, exactly, is it in cognitive development that's supposed to correspond to the emergence of early man's modular mind from the rudimentary general intelligence of the ancestral ape? Most cognitive psychologists who are seriously into cognitive modules think that modules are what children *start* with; certainly that's the natural view if you think that a modular intelligence is likely to be innate. And presumably it has to be innate if, as Mithen supposes, modularity is a product of the *evolution* of cognitive architecture.

Mithen relies heavily on some views of Karmiloff-Smith (1992; see chapter 10), who does indeed envision three phases of ontogeny, including both the assembly of rudimentary, task-specific capacities into modules, and transitions from modular intelligence to something that's supposed to be more fluid. But, precisely because she doesn't think modular architecture is genetically programmed, Karmiloff-Smith isn't by any means a paradigm modularity theorist. In effect, Mithen is trying so hard to construct an ontogenetic theory to ground phylogeny in, that he finds himself agreeing with people who don't really agree with one another—or with him.

Still Karmiloff-Smith might be right that ontogeny doesn't actually start with modules; or Mithen might be right about the prehistory of the mind even if the psychologists are wrong about its development. More worrying than whether Mithen can get phylogeny to run with ontogeny are the internal tensions in his account of the former.

For example, for Mithen's phylogenetic story to work, modules have to be smarter than generalized intelligence in order to account for the transition from the ancestral ape to early man; and cognitive fluidity has to be smarter than modules in order to account for the transition from early man to us. But, in fact, it seems that Mithen takes "generalized intelligence" and "cognitive fluidity" to be much the same sorts of things; indeed, he sometimes describes the phylogeny of mind as an "oscillation" from these to specialized intelligence and back again. But that can't be right; the same cause can't explain opposite effects. There must be something very different between, on the one hand, the kind of general intelligence that modular minds are supposed to have supplanted, and, on the other hand, the kind of fluid cognition that Mithen thinks allowed us to supplant Neanderthals. But Mithen doesn't say what this ingredient X might be. Nor does he say how it could be that, whereas modules emerge from general intelligence in the first transition, general intelligence emerges from modules in the second. Prima facie, one would think "emerging from" ought to be a one way street.

The trouble—and it's not just Mithen's trouble; it's a scandal in the whole cognitive science literature—is that nobody knows what "general intelligence" is or what "emerging from" is either. For example, there's a strong inclination to say that minds that are specialized must somehow evolve from minds that are less so; that sounds like good Darwinism. And there's also a strong inclination to say that it's because our intelligence is generalized and theirs is not that we're so much cleverer than other species. But, to put it mildly, it's not clear that you can have it both ways. I suspect that cognitive scientists want "general intelligence" and "emerging from" to paper over this crack, but that no coherent concepts could. No wonder if nobody can figure out what general intelligence and emerging from are.

Mithen does suggest that the transition from modularity to fluidity is somehow mediated by the use of language. Here again, he is explicitly echoing a literature in cognitive development according to which the transition from modular to generalized intelligence in children is linguistically mediated. But neither Mithen nor anybody else has been able to make clear how language manages to do the trick. The problem is that, if specialized concepts are the only ones that you have, then words that express specialized concepts are presumably the only ones that you can learn. In effect, the linguistic mediation story replaces the question "how do you get from specialized concepts to generalized concepts" with the question "how do you get from specialized concepts to *words for* generalized concepts." There's nothing to indicate that the second question is easier to answer than the first.

Well, any number can play this game. Here's my guess, for what it's worth, at how the pieces fit together. The phylogenetically primitive form of mind is not "generalized intelligence" but the reflex; that is, the more or less instantaneous, unmediated elicitation of an organic response by an impinging environmental stimulus. Two things have happened to reflexes in the course of the phylogeny of mind, both tending to decouple the stimulus from the response.

The first is the evolution of complicated, inferential processess of stimulus analysis, which parse an organism's world into a wide variety of behaviorally relevant categories; roughly, the higher the organism the more such categories its perceptual repertoire acknowledges. Reflexes respond to mechanical forces and to chemicals, light, gravity, and such; so unless these sorts of stimuli signal an opportunity for behavior, a mind that's made of reflexes is ipso facto blind to it. By contrast, our kinds of minds envision a world of objects, persons, and relations between them, upon the distribution of which the success of our behavior depends; so it's into these categories that we want perception to articulate the world that we act on. Phylogeny obliged, and apparently it started to do so early. Mechanisms for high-level perceptual analysis are much older than anything that could reasonably be called even rudimentary generalized intelligence. For example, insects use such mechanisms when they navigate (see Gallistel, 1990). Wherever they turn up, they exhibit the characteristic properties of modules: computational isolation, specialization, and genetic specificity. We probably have half a dozen or so of these modularized perceptual analyzers, as I remarked above.

Modules function to present the world to thought under descriptions that are germane to the success of behavior. But, of course, it's really the thinking that makes our minds special. Somehow, given an appropriately parsed perceptual environment, we often manage to figure out what's going on and what we ought to do about it. It's a hobbyhorse of mine that the thinking involved in this "figuring out" is really quite a different kind of mental process than the stimulus analysis that modules perform; and that it really is deeply mysterious. Suffice it that Mithen's model of phylogeny—that language somehow mediated a seepage of information from module to module—surely underestimates the difference between specialized intelligence and thought. Even if early man had modules for "natural intelligence" and "technical intelligence," he couldn't have become modern man just by adding what he knew about fires to what he knew about cows. The trick is in thinking out what happens when you put the two together; you get steak au poivre by integrating knowledge bases, not by merely summing them.

But "integration" is just a word, and we haven't a smidgen of a theory of what it is or how it happens.

On my view, the phylogeny of cognition is the interpolation, first of modularized stimulus analysis and then of the mechanisms of thought, between the Ss and Rs of aboriginal reflexive behavior. If that's right, then perhaps cognitive phylogeny hasn't much to learn from cognitive ontogeny after all. (Piaget taught that infants start out as sensory-motor reflex machines and develop into formal intelligences, but practically nobody outside Geneva believes that any more.) Quite possibly, the basic architecture of perception and thought are in place at birth, and what mostly happens to the child's mind in development is what also mostly happens to his body: Both get bigger.

I present this sketch as one kind of alternative to Mithen's picture. But, in fact, I wouldn't bet much on either. The synthesis of cognitive ontogeny with cognitive phylogeny is certainly premature; if somebody were to tell us the right story right now, we probably wouldn't understand it.

But then, synthesis has to be premature if it's to be of any use to guide research. We're not very good at thinking about minds; we need all the help we can get. Mithen has made a convincing case that "if you wish to know about the mind, do not ask only psychologists and philosophers: make sure you also ask an archaeologist." And he's drawn a first map of what might emerge from putting all their answers together. No doubt, we'll have to do the geography again and again before we get it right; but Mithen has made a start; more power to him.

Part IV

Philosophical Darwinism

Chapter 14

Review of Richard Dawkins's *Climbing Mount Improbable*

"How do you get to Carnegie Hall?" "—Practice, practice." Here's a different way: Start anywhere you like and take a step at random. If it's a step in the right direction, I'll say "warmer," in which case repeat the process from your new position. If I say "colder," go back a step and repeat from there. This is a kind of procedure that they call "hill climbing" in the computer-learning trade (hence, I suppose, the title of Dawkins's book). It's guaranteed to get you where you're going so long as the distance between you and where you're going is finite (and so long as there are no insurmountable obstacles or "local maxima" in the way. Nothing is perfect).

Hill climbing is often the theory of choice when a scientist's problem is to explain how something got to somewhere you wouldn't otherwise have expected it to be. That's in part because it's such an abstract and general sort of theory. All it requires is a source of random variation, a filter to select between the variants, and some "memory" mechanism to ensure that the selected variations accumulate. In all other respects, you're free to adapt it to whatever is worrying you.

For example, the "here" and "there" needn't be spatially defined. They might be, respectively, the undifferentiated primal, protoplasmic slime and the vast, intricate proliferation of species of organisms that now obtains. Darwinism (or, anyhow, the adaptationist part of Darwinism) is a hill-climbing account of the phenomenon of speciation: Genetic mutation takes the place of your random milling about; the inherited genetic endowments of successive generations of organisms correspond to the succession of positions that you occupy between here and Carnegie; and, instead of my shaping your path with gentle verbal cues, natural selection determines the direction of evolution by killing off mutations that happen to reduce organic fitness.

It all sounds pretty plausible. It might even be true. But the fact that a hill-climbing model could, in mathematical principle, find a way from where you started off to where you ended up, doesn't at all imply that you (or your species) actually got there that way. I could have just told

you where Carnegie Hall is, and you could have gotten there by following my instructions ("directed evolution"). Or I could have picked you
up and carried you ("interventionism"). Or you could have started out
at Carnegie, in which case the question wouldn't have arisen ("pre-
formation"). No doubt, other possibilities will occur to you. In the present
case, one might reasonably wonder whether we did, after all, get to be us
in the way that Darwinian adaptationism says that we did; it's reasonable
to want to see the evidence.

Especially so because the scientific success of the hill-climbing style of
explanation has often been underwhelming in other areas where it has
been tried. Classical economics (by which Darwin was apparently much
influenced) wanted to use it to account for the organization of markets. In
a system of exchange where gizmos are produced with randomly differing
efficiencies, canny consumers will filter for the gizmos that are best and
cheapest. Gizmos that are too expensive to buy or too cheap to sell at a
profit will automatically be screened out. Eventually an equilibrium will be
achieved that comports, as well as can be, with all the interests involved.

That's a nice story too. But, in the event, what often happens is that
the big gizmo-makers buy out the little gizmo-makers and supress their
patents. If there's still more than one gizmo-maker left in the field, they
compete marginally by painting their gizmos bright colors, or paying
some airhead to praise them on television. The evolution of gizmos
therefore grinds to a halt. Whichever producer a consumer decides to buy
his gizmos from, he finds that they don't work, or don't last, or cost too
much.

For another example, consider a version of hill-climbing theory that
used to be popular in psychology. How does a creature's behavior get
organized? How, for example, do you get from being a babbling baby to
being a fluent speaker of English? Here's how, according to B. F. Skinner
and the tradition of "reinforcement theory": Babbling is vocal behavior
that's produced at random. When you happen to make a noise that
sounds sort of like the local dialect, "society" reinforces you; and your
propensity to make that sort of sound (or better, your propensity to make
that sort of sound in those sorts of circumstances) increases correspondingly. Keep it up and soon you'll be able to say "Carnegie Hall" or "Jasha
Heifitz" or any other of the innumerable things that being able to speak
English allows you to say. Skinner used to complain to people who didn't
like his story about learning that he was just doing for the formation of
behavior what Darwin did for the formation of species. There was, I think,
some justice in that complaint, but it's an argument that cuts both ways.

In the event, language learning doesn't work by Skinnerian hill climbing. In particular, language learners don't make their errors at random in
the course of the acquisition process. Rather, as Noam Chomsky famously

pointed out, the grammatical and phonological hypotheses about language structure that children ever think to try out are sharply endogenously constrained. "Who Mummy love?" is recognizably baby talk; but "love Mummy who?" is not; it just isn't the kind of thing children say in the course of acquiring English. Ergo, it's not the kind of thing that society is ever required to filter out in the course of "shaping" the child's verbal behavior. But why isn't it, if children are hill climbing toward the mastery of English grammar, making mistakes at random as they go?

So there are at least two cases where, pretty clearly, applications of hill-climbing models tell less than all there is to be told about how a system gets organized. These examples have something striking in common. Hill climbing wants a random source of candidates to filter; but, in the market case and the language acquisition case, it appears that there are "hidden constraints" on which candidates for filtering ever get proposed. The market doesn't produce its gizmos at random, and the child doesn't produce its verbalizations at random either. The market is inhibited by restraint of trade, the child by (quite possibly innate) conditions on the kinds of language that human beings are able to learn and use. No doubt, in both cases, there is some residual random variability, and correspondingly some filtering which serves to smooth rough edges; so hill climbing gets a sort of vindication to that extent. But it's pyrrhic if, as practicioners in economics and psycholinguistics tend to suppose these days, it's the hidden constraints that are doing most of the work.

So the track record of hill-climbing explanations outside biology isn't what you'd call impeccable. What, then, about speciation? Nobody with any sense doubts that adaptation is part of the truth about evolution; but are there, maybe, "hidden constraints" at work here too? Or is the environmental filtering of random mutation most of what there is to how creatures evolve? Nobody loses absolutely all of the time. Maybe speciation is where hill climbing wins.

There is, in fact, currently something of a storm over just this issue, the vehemence of which Dawkins's book is much too inclined to understate. Paleontologists, since Darwin's own time, have often complained about what looks, from an anti-adaptationist perspective, like an embarassing lack of smooth gradations from species to species in the geological record. Maybe evolution gets from place to place by relatively big jumps ("saltations"), the intermediate options being ruled out by hidden contraints on what biological forms are possible. Something like this idea is at the heart of the current enthusiasm for evolution by "punctate equilibria." If you want to get to Carnegie, don't bother with exploring the intermediate loci; take a jet.

Dawkins doesn't make much of this sort of option; he's too busy assuring his lay audience that everything is perfectly fine chez classical adaptationism. Issues about evolution have become so politicized that a popularizing biologist must be tempted to make a policy of *pas devant les enfants*. Dawkins has, I think, succumbed a bit to this temptation. It's a disservice to the reader, who thereby misses much of the fun. For a corrective, try Niles Eldredge's 1995 book, *Reinventing Darwin*.

Anyhow, if classical adaptationism is true, then, at a minimum, the route from species A to its successor species B must be composed of viable intermediate forms which are of generally increasing fitness; there must be, in Dawkins's metaphor, smooth gradients leading up the hill that adaptation climbs. Much of Dawkins's book is devoted to an (admirable) attempt to make the case that there could have been such viable intermediates in the evolution of vision and of winged flight. Dawkins doesn't (and shouldn't) claim that any of these intermediate creatures are known to have existed. But he is pretty convincing that they might have existed, for all that biochemistry, physiology, embryology, and computer modeling have to tell us. The naive objection to adaptationism is that random mutation couldn't have made anything as intricate as an eye. Dawkins's answer is that sure it could; there's a physiologically possible path from bare sensitivity to light to the kind of visual system that we've got, and overall fitness would plausibly increase and accumulate as evolution traverses the path. It appears, in fact, that there may be several such paths; eyes have been independently reinvented many times in the course of evolution.

It is, however, one thing to show that evolution might have been mostly adapatation; it's another thing to show that it actually was. Many readers may be disappointed that Dawkins doesn't discuss at all the evolution of the piece of biology that they are likely to have most at heart: namely, human cognitive capacities. This is, on anybody's story, one of the places where the apparent lack of intermediate forms seems most glaring. Cognition is too soft to leave a paleontological record. And, pace some sentimental propaganda on behalf of chimpanzees and dolphins, there aren't any types of creatures currently around whose cognitive capacities look even remotely similar to ours. Moreover, there is a prima facie plausible argument that hidden constraints might play a special role in the evolution of a creature's psychological traits as opposed to, say, the evolution of its bodily form.

It's truistic that natural selection acts to filter genetic variation only insofar as the latter is expressed by corresponding alterations of a creature's relatively large-scale structure (alterations, for example, of the organs that mediate its internal economy or its environmental interactions). The slogan is: Genetic variants are selected for their *phenotypic* fitness. This holds, of course, for the case of nervous systems too; genetic endowments build neurological structures which natural selection accepts or rejects as it sees fit. Suppose that there is indeed relatively un-systematic variation not only in the genetic determinants of neurological structure, but also in the corresponding neurological phenotypes. Still, brain structures themselves are selected for the fitness of the psycho-logical capacities that they support. They're selected, one might say, not for their form but for their function. And nothing general—I mean *nothing* general—is known about the processess by which neurological alterations can occasion changes in psychological capacities.

Gradually lengthening the giraffe's neck should gradually increase its reach; that seems sufficiently transparent. But it's wide open what tumul-tuous saltations a gradual increase in (as it might be) brain size or the density of neural connections might cause in the evolving cognitive capacities of a species. The upshot is that *even if we knew for sure* that both genetic endowments and neurological phenotypes vary in a way that is more or less random and incremental, just as adaptationism requires, it wouldn't begin to follow that the variation of psychological traits or capacities is random and incremental too. As things now stand, it's per-fectly possible that unsystematic genetic variation results in correspond-ingly unsystematic alteration of neurophysiological phenotypes; but that the consequent *psychological* effects are neither incremental nor con-tinuous. For all anybody knows, our minds could have gotten here largely at a leap even if our brains did not. In fact, insofar as there is any evidence at all, it seems to suggest that reading brain structures onto mental capac-ities must substantially amplify neurological discontinuties. Our brains are, by any gross measure, physiologically quite similar to those of crea-tures whose minds are nonetheless, by any gross measure, unimaginably less intelligent.

Dawkins likes to "insist ... that wherever in nature there is a sufficiently powerful illusion of good design for some purpose, natural selection is the only known mechanism that can account for it" (202). He's right, I think, but this is another of those double-edged swords. The conclusion might be that adaptation really is most or all of what there is to evolution; or it might be that we don't actually know a lot about the etiology of what appears to be good biological design. Dawkins is inclined to bet on the first horse, but it's not hard to find quite reputable scientists who are inclined to bet on the second. Either way, it's a shame not to tell the reader that what's going on is, in fact, a horse race and not a triumphal procession.

Dawkins is the kind of scientist who disapproves of philosophy but can't stop himself from trying to do some. That's quite a familiar syn-drome. I should say a few words about what I'm afraid he takes to be the philosophical chapters of his book. They are, in my view, a lot less inter-esting than the biology.

Dawkins says, rightly, that Darwinism teaches us that the biological population of the world wasn't made for our comfort, amusement, or edification. "We need, for purposes of scientific understanding, to find a less human-centered view of the natural world." Right on. But then he spoils it by asking, in effect, if it's not all for us, who (or what) is it all for? This is a bad question, to which only bad answers are forthcoming.

The bad answer Dawkins offers in the present book follows the same line that he took in *The Selfish Gene*: It's all in aid of the DNA. "[W]hat are living things *really* [sic] for.... The answer is DNA. It is a profound and precise answer and the argument is watertight...." (247). The idea is that, from the gene's point of view, organisms are just "survival machines" whose purpose is to house and propogate the DNA that shaped them. A creature's only function in life (or in death, for that matter; see Dawkins's adaptationist treatment of the evolution of altruism) is to mediate the proliferation, down through the generations, of the genes that it carries. Likewise for the parts of creatures: "The peacock's beak, by picking up food that keeps the peacock alive, is a tool for indirectly spreading instructions for making peacock beaks" (252) (i.e., for spreading the peacock's DNA). It is, according to Dawkins, the preservation of the genetic instructions themselves that is the point of the operation.

But that doesn't work, since you could tell the story just as well from the point of view of any other of a creature's heritable traits; there's nothing special, in this respect, about its genetic endowment. For example, here is the Cycle of Generation as it appears from the point of view of the peacock's selfish beak: "Maybe genes think what beaks are for is to help make more genes, but what do they know about philosophy? Beaks see life steadily and they see life whole, and they think what genes are for is to help make more beaks. The apparatus—a survival machine, if that amuses you—works like this: beaks help to ensure the proliferation of peacocks, which help to ensure the proliferation of peacock DNA, which helps to ensure the proliferation of instructions to make more peacock beaks, *which helps to make more peacock beaks*. The beaks are the point; the beaks are what it's all 'for.' The rest is just mechanics."

What's wrong with this nonsense is that peacock beaks don't have points of view (or wants, or preferences), selfish or otherwise. And genes don't either (not even "unconsciously" though Dawkins is often confused between denying that evolutionary design is literally conscious and denying that it is literally design. It's the latter that's the issue). All that happens is this: microscopic variations cause macroscopic effects, as an indirect consequence of which sometimes the variants proliferate and sometimes they don't. That's all there is; there's a lot of "because" out there, but there isn't any "for."

In a certain sense, none of the teleological fooling around actually matters (which is, I guess, why Dawkins is prepared to indulge in it so freely). When you actually start to do the science, the metaphors drop out and statistics take over. So I wouldn't fuss about it except that, like Dawkins, I take science philosophically seriously; good science is as close as we ever get to the literal truth about how things are. I'm displeased with Dawkins's pop gloss on evolutionary theory because I think it gets in the way of seeing how science shows the world to be; and that, I would have thought, is what the popularizer of science-as-philosophy should most seek to convey. Dawkins is rather proud of his hardheadedness (he writes "sensitive" in sneer-quotes to show how tough he is); but in fact I think his naturalism doesn't go deep enough. Certainly it doesn't go as deep as Darwin's.

It's very hard to get this right because our penchant for teleology— for explaining things on the model of agents with beliefs, goals, and desires—is inveterate, and probably itself innate. We are forever wanting to know what things are for, and we don't like having to take Nothing for an answer. That gives us a wonderful head start on understanding the practical psychology of ourselves and our conspecifics; but it is one of the (no doubt many) respects in which we aren't kinds of creatures ideally equipped for doing natural science. Still I think that sometimes, out of the corner of an eye, "at the moment which is not of action or inaction," one can glimpse the true scientific vision; austere, tragic, alienated, and very beautiful. A world that isn't *for* anything; a world that is just there.

Chapter 15

Deconstructing Dennett's Darwin

Nobody likes me,
Everybody hates me,
I'm going to go out in the back yard and
Dig up some worms and
Eat them.
—Song we used to sing around the campfire when we were very young

Introduction

Dan Dennett's new book, *Darwin's Dangerous Idea*,[1] is full of anecdotes, digressions, thought experiments, homilies, "nifty" examples, and other erudite displays; with such a density of trees per unit forest, it's easy to lose one's way. But I think there is discernibly a main line to Dennett's argument. I think it goes like this: Dennett offers a sketch of a metaphysical construction in which the (neo-)Darwinian account of evolutionary adaptation is supposed to ground[2] a theory of natural teleology; and this theory of natural teleology is in turn supposed to ground an account of the meaning of symbols and of the intentionality of minds. This program is, of course, enormously ambitious; much more so than adaptationism per se. So you can without inconsistency buy the Darwin stuff but not the teleology; or you can buy the adaptationism and the teleology but not the theory of intentionality; or you can buy all three but doubt that Dennett has got it right about how they relate to one another.

My own view is that adaptationism probably isn't true; and that, even if it is true, there probably isn't any notion of natural teleology worth having; and that, even if adaptationism is true and there is a notion of natural teleology worth having, the latter isn't grounded in the former; and that even if adaptation grounds a theory of natural teleology, natural teleology has nothing much to do with the metaphysics of meaning.

It might be fun to explore all of that, but life is short and I have tickets for the opera. So, here's what I propose to do instead. I'll say just a little, mostly pro forma, about natural selection and natural teleology and the connections between them; then I'll suppress my qualms and grant, for the sake of argument, a sense of *selected for* and of *function* in which adaptations *can* be said to be selected for the function that they perform. (I think that what I'm thus proposing to concede, though it's pretty thin, is all the natural teleology that Dennett actually uses in his account of intentionality.) The main discussion will then concentrate on what I take Dennett to take to be the metaphysical connection between the teleological character of natural selection and the intentional character of thought. I'll argue, contra Dennett, that there isn't any such connection.

Adaptation

Dennett sometimes comes on pretty strong about the central status of the adaptationist paradigm in the overall scientific world view. "Adaptationist reasoning is not optional: it is the heart and soul of evolutionary biology. Although it may be supplemented, and its flaws repaired, to think of displacing it from central position in biology is to imagine not just the downfall of Darwinism but the collapse of modern biochemistry and all the life sciences and medicine" (238). That sounds pretty fierce, alright; if, like me, you're inclined to think that just about everything in this part of the woods is "optional," it should scare you half to death.

But it isn't clear that Dennett really means it. At one point, for example, he contemplates the possibility that Gould is right and there are "hidden constraints" on geneotypic variability (and/or its phenotypic expression) which may substantially reduce the space of options that natural selection ever gets to choose from: "The constraints of inherited form and developmental pathways may so channel any change that even though selection induces motion down permitted paths, the channel itself represents the primary determinant of evolutionary direction" (Gould, quoted by Dennett, 257). You might think that Dennett would reply that this would make the sky fall; but, in fact, he doesn't. His rejoinder is merely methodological: "[Even if] hidden constraints guarantee that there is a largely invisible set of maze walls ... in the space of apparent possibility ... we still can do no better in our exploration of this possibility than to play out our reverse-engineering strategies at every opportunity ..." (261). Or, as he puts it in a summary passage, "Good adaptationist thinking is always on the lookout for hidden constraints, and in fact is the best method for uncovering them" (261). That makes it look as though there is, after all, practically nothing that an adaptationist is required to believe about how evolution actually works; he's only required to buy into a methodological claim about how best to find out how it does. (Dennett doesn't, by the way, give any argument that assuming adaptationism is the "the best method" for uncovering hidden constraints; or even that it's a good method—as compared, e.g., with the direct investigation of the biophysics of genotypic variability. I don't blame him; it's notoriously hard to predict which way of doing a piece of science will work. The positivists were right about there not being a logic of discovery.)

So it's a bit unclear just what degree of commitment to adaptationism Dennett requires of biologists who wish to be in good scientific repute. It's also unclear exactly what doctrine it is that he requires them to endorse. Sometimes it seems pretty tepid: However it was we got here from the primal protoplasmic slime, no miracles ("skyhooks") were required on route. One might indeed take a hard line on that—I am myself prepared to sign on—since it is asymptotically close to empty. If we did come across a miracle, we'd call it a basic law or an anomaly and swallow it down accordingly. (Cf. electricity, action at a distance, and what goes on inside black holes.) Unlike many of my philosophical colleagues, I do think that there are substantial constraints that physicalism and reductionism impose on how we choose between special science theories. But these constraints are subtle and abstract. It's unimaginable that they should have the force of an a priori endorsement (or prohibition) of any substantive evolutionary theory.

Sometimes, however, what Dennett means by adaptationism isn't merely methodological and really does have teeth; it includes not just the familiar Darwinian schema *incremental, random variability → natural selection for fitness → inheritance,* but such very tendentious views as Dawkins's "selfish gene" story, according to which the Darwinian schema has primary applicability at the level of the individual gene (as opposed to, say, the unit of selection being the genotype, or the organism, or the breeding group, or the species). It's hard to imagine that Dennett really thinks that *that* theory is other than optional, or that the integrity of biology, or "the life sciences and medicine," or of "secular humanism" (see p. 476) turns upon it. But so his text sometimes suggests.

I propose not to get involved in this; it will do, for my purposes, merely to stipulate. So, let an adaptationist be, at a minimum, somebody who thinks that the explanation of many interesting phenotypic properties—and, in particular, of many of such phenotypic properties as distinguish species—invokes a history in which:

1. there is substantial and largely unsystematic genetic variation;
2. in consequence of which, there is substantial and largely unsystematic variation in the corresponding phenotypes;
3. natural selection operates differentially upon the phenotypic variants depending on their relative "fitness"; and
4. there is correspondingly differential inheritance of the genotypes of the selected forms.

I suppose something like this to be about as relaxed as (neo-)Darwinism can get short of lapsing into vacuity. For example, this formulation is neutral about what "fitness" consists in; it might be likelihood of individual survival, or likelihood of group survival, or likelihood of mating, or of contributing to the gene pool, or whatever. Also, the doctrine I've gestured toward isn't, as it were, "principled." One could hold that something like what it calls for really does happen some of the time, but that it doesn't happen all of the time; or that it's a part, but not the only part, of how evolution works; or that, though it's often true, it's wrong about the evolution of some interesting phenotypic properties of, for example, us. And so on.

I think that Dennett considerably overestimates the extent to which there is currently an adaptationist consensus in evolutionary theory (see the review by H. A. Orr, 1996). My own view is that Gould, Lewontin, Eldredge, and that crowd have made a substantial case that even this sort of relaxed Darwinism is right about a good deal less than the whole nature of evolution. I don't, however, propose to argue that this is so. For one thing, I have no claim at all to expertise in this area; and, for another, there's every indication that those guys can fight their own battles (see especially Eldredge, 1995). I can't, however, resist a word on why I'm particularly dubious about the application of even a relaxed Darwinism to the phenotypic properties that I care about most, namely, human cognitive traits and capacities.

Suppose that there is indeed a lot of relatively unsystematic genetic variability. It's truistic (and also not in dispute) that natural selection can act to filter this variability only insofar as it is expressed by corresponding phenotypic variation; genetic variants are selected for their *phenotypic* "fitness." Now, there are all sorts of ways in which genetic variation might turn out to be invisible at whatever level natural selection works on (individual phenotype; species phenotype; or what have you). And there are likewise all sorts of ways in which the magnitude of a genetic variation might underestimate the consequent alteration of a phenotype. For one thing, as Dennett rightly stresses, the phenotypic expression of a genotypic change is mediated by processes of (e.g.) protein synthesis that are, to put it mildly, complicated. It's thus open in principle, and also in empirical fact, how much genetic variability is dampened or amplified on the route to building an organism or a species: It can turn out that many genotypes converge on much the same phenotypes; or that only slightly different genotypes get grossly different phenotypic expression; or that identical genotypes get different phenotypic expressions in different contexts; etc. In short, it's perfectly possible that *even if genetic variation is unsystematic* (no "hidden constraints" at the genetic level) the processes that "read" it as phenotypic variation aren't. Thus Eldredge (op. cit.)

remarks that "[t]he implacable stability of species in the face of [underlying] genetic ferment is a marvelous demonstration that large-scale systems exhibit behaviors that do not mirror exactly the events and processes taking place among their parts...." (175)

All this holds, of course, for the psychological case inter alia; except that here it's one fillip worse since there is still another locus at which "hidden constraints" may have their effects. This point isn't novel, but it's well worth insisting on.

Suppose that there is relatively unsystematic variation in whatever are the genetic determinants of neurophysiological structure, and suppose further that this relatively unsystematic genetic variation is faithfully expressed by a correspondingly unsystematic variation of neurophysiological phenotypes. Still, the selection of these neurophysiological phenotypes must itself be mediated by whatever processes govern their expression as psychological traits and capacities.[3] And nothing general— I mean *nothing* general—is known about these processes. The upshot is that *even if we knew for sure* that both genetic and neurophysiological variability are more or less incremental, random, and unsystematic, it wouldn't begin to follow that the variability of psychological traits or capacities is too.

So, for example, as Chomsky and Gould have frequently remarked, it is perfectly possible that there is random variability of, and environmental selection for, overall brain size, and that this is as gradualistic and classically Darwinian as you like. That would still leave it open that, at some point(s) or other, and by processes currently utterly unknown, the adventitious consequences of such physiological change are radical *discontinuities* in behavioral repertoires. (Or, if you don't like brain size, try brain weight, or brain/body ratio, or the density of neural connections, or the amount of surface folding, or the proportion of cortex to old brain, or the proportion of neurons to glia or, for that matter, the color of the damned thing. Take your pick. All except the last have been in fashion, at one time or another, as candidates for conveying selectional advantage.) I repeat, *nothing* general is known about how physiological variation determines variation of psychological traits and capacities. Do not believe anything to the contrary that you may have read in the Tuesday *New York Times*.

To summarize this line of thought: It is, of course, an empirical issue how close to being true even a relaxed adaptationism will prove to be. For it to be very close, the effects of hidden constraints on the course of evolution have to be relatively small compared to the effects of unsystematic variation of genotypes and their phenotypic expressions. In the psychological case, however, the notion of "phenotypic expression" is ambiguous; it may refer to a creature's neurophysiology or to its behavioral traits and capacities. So there are not just one but *two* places where hidden constraints may enter into determining the evolutionary process. Even if the *physiological expression of geneotypic variation* is largely unconstrained, there is, as things now stand, no reason to suppose that the *behavioral expression of physiological variation* is too. In fact, insofar as there is any evidence at all, it seems to suggest that reading neural structures onto behavioral repertoires must substantially amplify neurophysiological discontinuities. Our brains are, by any gross measure, physiologically quite similar to those of creatures whose psychological capacities are nonetheless, by any gross measure, unimaginably inferior.

Pace Dennett, even a relaxed adaptationism about our psychological traits and capacities isn't an article of scientific faith or dogma; we'll just have to wait and see how, and whether, our minds evolved. As of this writing, *the data aren't in.*

Adaptation and Teleology

Suppose, however, that adaptationism is true; is it able to ground a notion of natural teleology? Or, to put it in terms a little closer to Dennett's, suppose there is a type of organism O that has some genotypic property G that was selected because G's (characteristic) phenotypic expression P increases O's relative fitness (on average, ceteris paribus, in ecologically normal circumstances ... blah, blah, blah). Is it then reasonable to speak of P as a property that O was "designed" to have? Or as a "solution" to an "engineering problem" that O's ecology posed?

It's important to see how this kind of question might bear on issues about intentionality. If adaptations are, in some not entirely figurative sense, design features of organisms, then maybe the root metaphysical problem about how to get meaning into a purely mechanical world order is on its way to being solved. Just how this might go will concern us presently. It will do, for now, to notice that DESIGNED FOR F-ING, and the like, belong to the same family of concepts as DESIGNED WITH F-ING IN MIND. Perhaps, then, the putative connection between natural selection and design is the thin end of the intentional iceberg. Perhaps there is a route from the indubitably mechanistic adaptationist story about how phenotypes evolve, to the intentional idiom which (eliminativists excepted) everybody thinks that a good theory of the mind requires. That, in a nutshell, is how Dennett is hoping that things will pan out.

The subtext is the thing to keep your eye on here. It is, no doubt, an interesting question in its own right whether adaptationism licenses teleological notions like SELECTION FOR. But what makes that question interesting in the present metaphysical context is that SELECTION FOR is presumably *intensional.*[4] Just as you can believe that P and not believe that Q even though P iff Q, so a creature might be selected for being F

and not for being G even though all and only the Fs are Gs. Correspond-ingly, the issue we're most concerned with is whether a naturalistically grounded notion of *selection for* would be intensional in the ways, and for the reasons, that mental states are. If so, then maybe a naturalistic teleol-ogy is indeed a first step toward a naturalistic theory of mind.

But, promising though it may seem, I'm afraid this line is hopeless, and for familiar reasons. Design (as opposed to mere order) requires a designer. Not theologically or metaphysically (pace Paley, Bishop Wilberforce et al.), but just conceptually. You can't explain intentionality by appealing to the notion of design because the notion of design *presupposes* intention-ality. I do think this is obvious and dreary and that Dennett should give up trying to swim upstream against it (especially since, as we're about to see, there's a different route to what he wants that will probably suit his purpose just as well).

Anyhow, here's the (familiar) argument. Patently, not every effect that a process has is ipso facto an effect that it designs; short of theology, at least some effects of every process are merely adventitious. This must hold of the process of natural selection inter alia. So, in evolutionary theory as elsewhere, if you wish to deploy the idiom of posed problems and designed solutions, you must say something about what designing requires *over and above mere causing*. Lacking this distinction, everything a process causes is (vacuously) one of its designed effects, and every one of its effects is (vacuously) the solution to the problem of causing one of *those*.

To be sure, if solutions aren't distinguished from mere effects, it does comes out—as Dennett would want it to—that the giraffe's long neck solved the problem of reaching to the top of things, and did so under precisely the ecological conditions that giraffes evolved in. But equally, and in exactly the same sense, I solve the problem of how to make a Jerry Fodor under the genetic and environmental conditions that obtained in making me; and the Rockies solve the problem of how to make mountains *just like the Rockies* out of just the materials that the Rockies are made of and under just the conditions of upthrust and erosion in which they formed; and the Pacific Ocean solves the problem of how to make a hole of just that size and just that shape that is filled with just that much salt water; and the tree in my garden solves the problem of how to cast a shadow just that long at just this time of the day.[5] This is, however, no metaphysical breakthrough, it's just a rather pointless way of talking; nei-ther I nor the Pacific get any kudos for being solutions in this attenuated sense. That's because problems are like headaches; they don't float free of people's having them. The Pacific and I didn't *really* solve anything because *nobody had* the problems that we would have been solutions to. ("Who would want *those?*" people always ask.)

Serious talk about problems and solutions requires a serious account of the difference between designing and merely causing. Notice, moreover, that if your goal is a reductive theory of intentionality, then your account of this difference cannot itself invoke intentional idiom in any essential way. This really does make things hard for Dennett. In the usual case, we distinguish designing from mere causing by reference to *the effects that the designer did or didn't intend.* For example: The flowers Sam gave Mary made her wheeze and did not please her. They were, nonetheless, a failed solution to the please-Mary problem, not a successful solution to the wheeze-Mary problem. That's because Sam intended that receiving the flowers should please her and did not intend that they should excite her asthma. Suppose, by contrast, that Mary merely *came across* the flowers, and that they both pleased her and made her wheeze. Then the flowers didn't solve, or fail to solve, *anything;* they just had whatever effects they did. I think the intuitions here are about as clear as intuitions can be. It certainly looks as though the concept of design presupposes, and hence *cannot be invoked to explain,* the accessibility of intentional idiom.

If you found a watch on a desert island, you'd have a couple of options. You could argue that since it was clearly designed, there has to have been a designer; or you could argue that since there was certainly no designer, the watch can't have been designed. What is *not,* however, available is the course that Dennett appears to be embarked upon: there was no designer, but the watch was designed all the same. *That just makes no sense.*

There may be a way out of this somewhere in Dennett's present text or elsewhere in his writings; but if there is I honestly can't find it. I think perhaps Dennett is led to underestimate the magnitude of this problem because he allows himself such free use of the metaphor of Mother Nature as the designer from whose planning the intentionality of selection flows. But surely none of that talk can be meant to be literal; reduction has to stop somewhere, but it can't conceivably stop *there.* "Jerry Fodor may joke about the preposterous idea of our being Mother Nature's artifacts, but the laughter rings hollow; the only alternative views posit one sky-hook or another" (427). But Dennett doesn't seriously think that having a scientific world view requires believing *that,* does he? The story about Mother Nature is, after all, a fairy tale. There *isn't* any Mother Nature, and no intentional agent literally planned me or designed the Rockies or saw to the tree in the garden. Could Dennett really have lost sight of that?[6]

You can't get natural teleology by postulating designerless designs. Still, both for the sake of the argument and because it may be true, I'm prepared to grant that a different approach to metaphysical reduction really might yield a certain kind of natural teleology. Consider a well-worn example (e.g. Hempel, 1965). Hearts pump blood and they make the familiar sort of noise, and this correlation is empirically reliable. Nonetheless, intuition strongly suggests that hearts were selected for being blood pumps, not for being noise makers. That is, in such cases intuition reads "selected for" as *intensional*:[7] The inference from "*x*s were selected for *F*" and "(reliably) *F* iff *G*" to "*x*s were selected for *G*" is *not* valid. What, if not appeals to the intentions of a designer, underwrites such intuitions?

Maybe appeals to counterfactuals do. Maybe, in the present case, the counterfactual that legitimizes the inference is that hearts *would* contribute to fitness (hence would be selected) in (nearly?) worlds where they pump blood but don't make noise; but they wouldn't contribute to fitness (hence wouldn't be selected) in (ditto) worlds where they make noise but don't pump blood. At least this way of getting intensionality into the picture doesn't invoke the operation of intentional systems (minds);[8] to that extent it might be of use to a reductivist metaphysical program.

This kind of approach to natural teleology has well-known problems, to be sure. For one thing, it's not entirely clear that the required counterfactuals are true. Teleological explanations generally offer sufficient but not necessary conditions for selection. Maybe if hearts didn't pump, they would be selected for something else. (Hempel, op. cit.; Dennett recognizes this sort of point in a slightly different context. See p. 259–260).

For another thing, maybe there just aren't any (nearby?) worlds of the kind that the counterfactual contemplates. It's arguably *a law*—hence nomologically necessary—that hearts pump blood iff they make noise; and who knows what to make of counterfactuals whose antecedents are *necessarily* false. Not only: "who knows what their truth conditions, if any, are?" but also "who knows what, if any, roles they can play in empirical, for example, biological, theorizing?"

Digression: Ruth Millikan (1984) has been pushing a line rather different from Dennett's (if I read her right; which, however, Dennett says I hardly ever do) according to which what a trait is selected for are those of its effects that (completely) explain why a creature has it. Correspondingly, the intensionality of "selected for" derives from the intensionality of "(completely) explains." (Presumably from *x*s being *F* [completely] explains *y*'s being *G* and *x*s are *F* iff *x*s are *H*, it does *not* follow that *x*s being *H* [completely] explains *y*'s being *G*.)

I am pretty comprehensively unmoved by this. Though it's true that mere extensional equivalence does not license the substitution of "*F*" for "*G*" in *F explains that* ..., it looks like (nomologically) *necessary* equivalence does. (See fn. 4.) In particular, if "*F* iff *G*" is a law, then for every explanation according to which trait *T* is selected for because it causes *F*, there will be a corresponding, equally complete, equally warranted explanation according to which *T* is selected for because it causes *G in a world where, reliably, Gs are Fs.*

The standard philosophical illustration goes like this: Is what explains the fitness of the frog's feeding reflex that it causes snaps at flies? Or is what explains its fitness that it causes snaps at ambient black dots in circumstances where it is nomologically reliable that the ambient black dots are flies? Unless there is something to choose between these explanations, there will be nothing to choose between the corresponding teleological claims about what the behavioral repertoires of frogs are for. Notice that, though this example concerns the selection of an intentional trait, the problem it invokes is *entirely general*: Appeals to being F can explain nothing that isn't equally well explained by appeals to being G *in a world where it's necessary that Gs are F*.[9] In effect, contexts of explanation are transparent to the substitution of (e.g., nomologically) necessary equivalents. (For a less sketchy discussion of this group of issues, see Fodor, 1990 and papers in Loewer and Rey, 1991.)

As remarked above, in the present context the main issue is whether a naturalistic reduction can license *intensional* teleological idiom. Here, then, is what's to choose between Millikan's and Dennett's account of *selection for*. Dennett *can* distinguish between selection for nomologically equivalent F and G. That's because he helps himself to "designed for"; and a thing can be designed for F-ing and not for G-ing even if it's a law that only Gs F; designing is *more* intensional than explaining. The down side, for Dennett, is that there is no design without a designer; so, willy nilly, he finds himself in bed with Mother Nature. Millikan tries rather harder not to cheat; *explains* is (arguably) a relation between *propositions*, so (maybe) it doesn't tacitly invoke an agent.[10] The down side, for Millikan, is that explanation isn't intensional *enough*; unlike the intensionality of the mental (and, in particular, unlike the intensionality of design), explanation doesn't distinguish necessary equivalents. And so neither, of course, does a notion of *selection for* that "explains" is used to explain.

Much of a not muchness, if you ask me.

Well, so much about several approaches to natural teleology, and the qualms that they inspire. I'm sort of inclined to doubt that there *is* any natural teleology; I sort of think that the only goals there are the ones that minds entertain. But never mind. For the sake of the argument, I hereby grant that there is such a thing as *selection for* after all, that it can be naturalistically grounded, and that selection for F doesn't imply selection for G even if "F iff G" is reliable. The argument will now be that, even so, the intensionality of "selection for" doesn't help with the metaphysics of the intentionality of minds.

Deconstruction

I'm proposing to deconstruct what I take to be Dennett's metaphysics of intentionality. Now, as everybody knows who's in this line of work, the way you set about a bit of deconstructing is: You winkle out of the text some aporia (that means, roughly, *glitch*, or *screw-up*) that reveals its deep, unconscious yen for self-refutation. (Texts are what have yens these days, the author being dead.) It is in this fashionably postmodern spirit that I call your attention to what strikes me as distinctly a peculiarity in Dennett's views.

On the one hand, it's of the essence of Dennett's program to argue that there is no *principled* difference between the intentionality of natural selection and the intentionality of mind. A *principled* difference would need a skyhook to get over it, whereas Dennett's primary effort is to overcome the "resistance philosophers have shown to evolutionary theories of meaning, theories that purport to discern that the meaning of words, and all the mental states that somehow lie behind them, is grounded ultimately in the rich earth of biological function" (402). Or, as he puts it in a summary passage: "Real meaning, the sort of meaning our words and ideas have, is itself an emergent product of originally meaningless processes—the algorithmic processes that have created the entire biosphere" (402).

It is because he cares so much about the continuity of the intentionality of "words and ideas" with the intentionality of selection that Dennett is so emphatic in denying that the distinction between "derived" and "original" intentionality is principled. Since "... your selfish genes can be seen to be the original *source* of your intentionality—and hence of every meaning you can ever contemplate or conjure up," the soi-disant "original" intentionality of your thoughts is every bit as derived as the patently parasitic intentionality of thermostats and street signs. Thermostats and street signs derive their intentionality from us; we derive our intentionality from Mother Nature (specifically, from the reproductive ambitions of our selfish genes). Only the evolutionary process is originally intentional *sans phrase*.

That is one part of Dennett's story. But there is another part, equally strongly insisted on, that might well strike one as distinctly in tension with the sorts of passages just cited. For, it turns out that "... Mother Nature (natural selection) can be viewed as having intentions, [only] *in the limited sense of having retrospectively endorsed features for one reason or another* ..." (462) and "The Local Rule is fundamental to Darwinism; it is equivalent to the requirement that there cannot be any intelligent (or farseeing) foresight in the design process, but only ultimately stupid opportunistic exploitation of whatever luck ... happens your way" (191). In this respect, it appears, Mother Nature is *not*, after all, much like you and me. "If [for example] you were playing chess under hidden constraints, you would adjust your strategy accordingly. Knowing that you had secretly promised not to move your queen diagonally, you would probably forgo any campaign that put your queen at risk of capture ... but you have knowledge of the hidden constraints and foresight. Mother Nature does not. Mother Nature has no reason to avoid high-risk gambits; she takes them all, and shrugs when most of them loose" (259). It is, indeed, precisely because of her lack of foresight that Mother Nature is forever getting trapped in local maxima: "If only those redwoods could get together and agree on some sensible zoning restrictions and stop competing with each other for sunlight, they could avoid the trouble of building those ridiculous and expensive trunks" (255; Dennett is quoting Dennett 1990, 132).

So, on the one hand, "There may be somewhat nonarbitrary dividing lines to be drawn between biological, psychological, and cultural manifestations of this structure [decision making by 'satisficing'], but not only are the structures—and their powers and vulnerabilities—basically the same, the particular contents of "deliberation" are probably not locked into any one level in the overall process but can migrate" (504). But, on the other hand, Mother Nature is a *blind* watchmaker; and her blindness consists (inter alia, see below) in this: Her reasoning is never *prudential*; she maximizes her utilities *only in retrospect*; she *never* adopts (or rejects) a policy because of outcomes that she has *foreseen*—whereas, you and I do that sort of thing all the time. *Why doesn't this difference between her and us count as principled?* I call that an aporia (or do I mean an aporium? Probably not).

None of this, please note, takes back so much as a word of what I've agreed to concede. In particular, I am still reading "select for" and the like as intensional. But though by assumption Mother Nature's *selecting for ...* and my *intending to ...* have their intensionality in common, I'm now wanting to emphasize a residual difference between them. Mother Nature never rejects a trait because she can imagine a more desirable alternative, or ever selects for one because she can't. We do.

Is this difference principled? I'll argue that it is, anyhow, qualitative; it shows that the intentionality of minds is a different kind of thing from the (putative) intentionality of selection. First, however, it's important to get clear on the scope of the problem. What we've got so far is only the entering wedge.

It's true but a bit misleading to say that Mother Nature can't foresee the consequences of selection. For, if she can't see forward, she can't see sideways or backwards either; all she can see is where she is. So, just as Mother Nature never prefers F to G because it will (likely) bring it about that H, so she also never prefers F to G because it *would* bring about that H if, counterfactually, it *were* the case that K; and she never prefers F to G because Fs but not Gs used to be Js (though they're not anymore). In short, it's not the (likely) future as such that Mother Nature is blind to; it's the *nonactual as such* that she can't see.

In fact, even that underestimates the magnitude of her myopia. Just as the process of selection is blind to whatever is unactual, so too is it unaffected by any actual thing that is, de facto, causally isolated from the process of selecting. Hence, for example, one aspect of the impact of geography upon evolution: Let it be that Fs are predators on Gs; and let it be that some genetic variation produces a strain of G*s that are invisible to Fs. Still, being F-invisible won't provide these G* variants with any selectional advantages over the old fashioned sort of Gs if, as it happens, *all the Fs live in some other part of the forest.* Selectional advantage is the product of, and only of, *real causal interactions.* Merely possible competitions don't enter into Mother Nature's calculations; not even if they are possible competitions between *actual* creatures. None of this, surely, is in the least contentious; in fact, it's close to tautology: *Nothing is ever the effect of merely possible causes. Nothing is. Not ever.*

Mother Nature never prefers any Fs to any Gs *except on grounds of (direct or indirect) causal interactions that the Fs and G actually have with the selection process.* This looks to me like a whopping difference between Mother Nature and us. Indeed, if you think of intentionality in the Good Old Brentano Way, it is exactly what makes mental states *intentional* and not just *intensional,* namely, their capacity for having nonactual objects.[11]

Here's a way to summarize the point: Mental states are *both* intensional *and* intentional. Or, if you prefer, mental state predicates resist *both* substitution of coextensives *and* quantifying in. So, in the first case, Lois may fail to realize that Clark is Kent even though (ho hum!) she knows quite well, thank you, that Clark is Clark. And, in the second case, Lois may believe that Santa Claus comes down the chimney even though (ho hum!) there isn't any Santa Claus. Now, I don't really think that the intensionality of selection explains the intensionality-with-an-s of mental states. That's because, as I remarked above, however its intensionality is grounded, "selected for" is likely to be transparent to the substitution of *necessarily* coextensive terms, whereas "realizes," "thinks about," "believes that," and the like certainly are not. But even if selection did explain why the mental is intensional-with-an-s, it *still* wouldn't explain why the mental is intentional-with-a-t. That's because a state that is intentional-with-a-t can have "Ideal" or "intentionally inexistent" objects, and its causal role can (typically does) depend on which Ideal object it has. Whereas, to repeat, only what is *actual* is visible to Mother Nature. *Only* what is *actual* can affect the course of a creature's selectional history.

This sort of point is occasionally remarked upon in the literature on natural teleology; see, for example, Allen and Bekoff (1995): "For a thing to possess a biological function, at least some (earlier) members of the class must have successfully performed the function. *In cases of psychological design, the corresponding claim is not true*" (61; my emphasis). That is:

unlike the traits that Mother Nature selects (for), the goals to which
intentional activity is directed can be entirely Ideal. Philosophers whose
project is the reduction of intentionality to natural teleology seem not to
have understood how much difference this difference makes.

But, you might reasonably wish to argue, this can't be a problem for
Dennett's kind of reductionism unless it's a problem for *everybody's* kind of
reductionism. For, you might reasonably wish to continue to argue, if
nothing can be the effect of a merely Ideal cause, then, a fortiori, thoughts,
decisions, and actions can't be the effects of merely Ideal causes. And similarly for other intentional goings on. Remember that Brentano thought
that the intentionality of the attitudes shows that naturalism can't be true;
and Quine thinks that naturalism shows that the attitudes can't be intentional; and the Churchlands think (at least from time to time) that the
intentionality of the attitudes shows that there can't be any attitudes. If, in
short, there's an argument that Mother Nature is blind to merely possible
outcomes, then there must be the same argument that you and I are blind
to merely possible outcomes. So either Dennett wins or everybody
looses.

Now, I do think that's a puzzle; but it's not one that I just made up. In
fact, it's just the old puzzle about intentionality-with-a-t (as distinct from
the old puzzle about intensionality-with-an-s). How can what is only Ideal
affect cognitive and behavioral processes? How can I think about, long
for, try to find, blah, blah, the gold mountain *even though there isn't any
gold mountain?* This is a question which, though it arises for mental processes, *has no counterpart for processes of natural selection* since, to repeat,
although "selected for" and the like are maybe sometimes opaque to the
substitution of equivalents (viz., when the equivalence is not—nomologically or otherwise—necessary), they are patently always transparent to
existential quantification.

Preliminary moral: You cannot solve the problem of meaning by reduc-
ing the intentionality of mind to the intensionality of selection, because
the intensionality of selection fails to exhibit the very property that
makes meaning problematic. Mother Nature has no foresight, so she can't
ever select for a trait that isn't there. But *you can mean something that isn't
there;* you can do that any time you feel like.

But then, what sort of story *should* one tell about intentionality-with-
a-t? There's a commonsense approach which, in my view, probably points
in the right direction but which, no doubt, Dennett would think invokes a
skyhook: Merely intentional objects can affect the outcomes of cognitive
processes *qua, but only qua, represented.* What *appear* to be the effects of
Ideal objects on causal processes are invariably mediated by *representations*
of these objects; and *representations*, unlike their *representees*, are actual by
stipulation (they're physical tokens of representation types). The fact that

elephants fly can't make anything happen because there isn't any such fact. But tokens of the symbol type "elephants fly" mean that elephants fly, and they can make things happen because there are as many actual ones of those as you like.

That doesn't, of course, *solve* the problem of intentionality; it merely replaces it with the *unsolved* problem of representation (i.e., of "meaning that"). Now, I don't know whether representation is a skyhook, *and neither does Dennett*. Either there is a naturalistic theory of representation—in which case, it *is* the solution to the problem of intentionality—or there is no naturalistic theory of representation, in which case I, for one, will give it all up, become an eliminativist about the mental, and opt for early retirement (well, early*ish*). But, either way, the present point stands: You can't reduce intentionality to "selection for" because *selection for doesn't involve representation.* (A fortiori, selection for *F*-ness doesn't involve representing anything as *F*.) That, in a nutshell, is the difference between what Mother Nature does when she *selects for* tall petunias, and what Granny does when she *breeds for* them.[12]

You can, in short, suppose that the whole (neo-)Darwinian story is true; and you can suppose that "selection for" is intensional; but you will not thereby have succeeded in supposing any representation into the world. And, according to common sense (and according to me), it's representation that you need to explain intentionality.

Three caveats and we're finished. First, I haven't claimed that only minds can represent. On the contrary, us informational semanticists are inclined to think that the representational is a (possibly large) superset of the intentional. So, for example, it's okay with us if genes, or tree-rings, or the smoke that means fire have (underived) representational content. It's just that, according to the previous argument, they don't have it *qua selected.*

Second, I hope you don't think that I think that the line of argument I've been pursuing shows that natural selection couldn't have *resulted in* intentional processes. Of course it could; or, anyhow, of course it could for all that I know. The issue about the "source" of intentionality is not the *historical* question "what made intentional things," but the *metaphysical* question "what makes intentional things intentional." If what produced intentionality turns out to have been the action of electricity on the primal soup, it wouldn't follow that intentionality is either a kind of soup or a kind of electricity.

Third, a point that should be obvious: It's okay for somebody who is a reductionist or a naturalist about intentionality to not believe that intentionality reduces to adaptation or natural teleology. Dennett, having appropriated "naturalistic" for his own brand of reduction, seems unable to contemplate a view that says "yes, intentionality is reducible, but no

it's not reducible to Darwin." Hence such really bizarre misreadings as "...
Searle and Fodor ... concede the possibility of [a clever robot] but ... dispute its "metaphysical status"; however adroitly it managed its affairs,
they say, its intentionality would not be the real thing" (426).

But I don't say that, and I never did (and, as I read him, Searle doesn't
and didn't either. I may have mentioned that at one point, Dennett chas-
tises me for being unsympathetic in my reading of him and Millikan. Talk
about your pots and your kettles!) What I have said and still say is that,
however clever a robot (or a creature) is at managing its affairs, its intentionality, its *having affairs to manage*, does not *consist in* the cleverness
with which it manages them. If it did, intentionality would supervene on
behavior and behaviorism would be true—which it's certainly not.

I also say that, however clever a robot (or a creature) may be, and
however much its cleverness may, in point of historical fact, have been
the cause of its having been selected, its intentionality does not *consist in*
its having been selected for its cleverness. Dennett really must see that
saying these sorts of things is quite compatible with being as naturalist
and reductionist as you like about minds. "Jerry Fodor may joke about the
preposterous idea of our being Mother Nature's artifacts, but the laughter
rings hollow; *the only alternative views posit one skyhook or another.*" Nonsense. There are lots of alternatives to adaptationist accounts of intentionality. Some are eliminative and some are reductive; some are naturalistic
and some aren't; some are emergentist and there are even one or two
that are panpsychist. What they have in common is only their being
reconciled to a thought that Dennett apparently can't bring himself to
face, a truly Darwinian idea, and one that is, if not precisely dangerous, at
least pretty disconcerting: *There isn't any Mother Nature*, so it can't be that
we are her children or her artifacts, or that our intentionality derives from
hers.

Dennett thinks that Darwin killed God. In fact, God was dead a century
or so before Darwin turned up. What really happened was that the
Romantics tried to console themselves for God's being dead by anthro-
pomorphizing the natural order. But Darwin made it crystal clear that the
natural order couldn't care less. *It wasn't God that Darwin killed, it was
Mother Nature.*

Notes

1. Readers who are puzzled by the epigraph should consult p. 17 of Dennett (1995). All
    Dennett references are to that volume except as indicated.
2. What's this "grounding"? Search me; but philosophers, including Dennett, often like to
    talk that way. For present purposes, theory A "grounds" theory B only if (i) the ontol-
    ogy of A reduces the ontology of B, and (ii) A is independently justified.

3. Likewise, the organism's phenotypic constellation of psychological traits and capacities is filtered and transformed by whatever processes "read" it onto the creature's behavior. Hence the need for a "performance/competence" distinction *within* theories at the psychological level.

4. See Sober (1984), whose generally admirable treatment of these issues is, however, mistaken at a crucial point. Sober claims that *selection for*—but not *selection*—is intensional, and also that what gets *selected* is objects, whereas what gets *selected for* is traits (properties). But, as we're about to see, *selection for* contexts are *extensional* (viz., transparent, like *selection* contexts) in respect of the substitutivity of coextensive trait terms when the coextension is (e.g., nomically) *necessary*. That, in fact, is one of the main reasons that *selection for* is so unlikely to ground a theory of the intensionality of the mental.

5. Moreover, all these solutions meet Dennett's requirement of being "algorithmic." It is, however, unsurprising that they do so since, as Dennett rightly says, there aren't "any limits at all on what may be considered an algorithmic process...." (59) Step right up and play; every contestant is a guaranteed winner.

6. It adds to the confusion that Dennett often writes as though "Mother Nature" is just a figurative name for natural selection. That is, of course, perfectly okay; but you can't *both* talk that way and also hold, as a substantive thesis, that *her* intentionality explains *its*.

7. The wary reader will notice that I have switched from intentionality-with-a-t to intensionality-with-an-s. Quite so; the point of this will appear in due course.

8. Unless the truth makers for counterfactuals are themselves mind-dependent, as, indeed, philosophers have sometimes supposed.

9. It's unclear how much of this sort of teleological or intensional indeterminacy Millikan is prepared to live with. Dennett, however, is explicit that he doesn't mind if there is no fact of the matter about the intentional object of the frog's snap. But he doesn't say whether he's equally sanguine about *every* ascription of intentional content to the frog being indeterminate between reliably coextensive properties. Nor does he say why, on his view, the same wouldn't hold of intentional ascriptions to *us*. (Do not reply "that's because we have language" or I shall break down and commence to gnaw the rug: Exactly the same indeterminacy would infect the assignment of meanings to reliably coextensive predicates of, say, English.)

10. I'm being very nice and assuming, concessively, that explanation isn't itself a pragmatic notion. If what explains what depends on who wants to know what, then Millikan's line is just as question-begging as Dennett's.

11. This last is by way of cashing footnote 7.

12. Sober (op. cit., p. 202) remarks that "Artificial selection is a variety of natural selection; the relation is one of set inclusion, not disjointness." Well, yes and no. It's quite true that, if you are typing selectors by the *effects* of their sorting, Granny and Mother Nature are both just filters on phenotypic variance. But if you are typing selectors by *how they achieve their effects*, the difference could hardly be more striking. In the whole natural order, as far as anybody knows, only Granny and her kind filter with, literally, *premeditation*.

Chapter 16

Is Science Biologically Possible? Comments on Some Arguments of Patricia Churchland and of Alvin Plantinga

I hold to a philosophical view that, for want of a better term, I'll call by one that is usually taken to be pejorative: *Scientism*. Scientism claims, on the one hand, that the goals of scientific inquiry include the discovery of objective empirical truths; and, on the other hand, that science comes pretty close to achieving this goal at least from time to time. The molecular theory of gasses is, I suppose, a plausible example of achieving it in physics; so is the cell theory in biology; the theory, in geology, that the earth is very old; and the theory, in astronomy, that the stars are very far away. Scientism is, to borrow a phrase from Hilary Putnam the scientist's philosophy of science. It holds that scientists are trying to do pretty much what they say that they are trying to do; and that, with some frequency, they succeed.

I'm inclined to think that scientism, so construed, is not just true but *obviously and certainly* true; it's something that nobody in the late twentieth century who has a claim to an adequate education and a minimum of common sense should doubt. In fact, however, scientism is tendentious. It's under attack, on the left, from a spectrum of relativists and pragmatists, and, on the right, from a spectrum of Idealists and a priorists. People who are otherwise hardly on speaking terms—feminists and fundamentalists, for example—are thus often unanimous and vehement in rejecting scientism. But though the rejection of scientism makes odd bedfellows, it somehow manages to make them in very substantial numbers. I find it, as I say, hard to understand why that is so, and I suppose the Enlightenment must be turning in its grave. Still, over the years I've sort of gotten used to it.

Recently, however, the discussion has taken a turn that seems to me bizarre. Scientism is now being attacked by, of all people, *Darwinists*. Not all Darwinists, to be sure; or even, I should think, a near majority. Still, the following rumor is definitely abroad in the philosophical community: When applied to the evolution of cognition, the theory of natural selection somehow entails, or at a minimum strongly suggests, that most of our empirical beliefs aren't true; a fortiori, that most of our empirical scientific theories aren't true either. So the rumor is that Darwinism—which, after all, is widely advertised as a *paradigm* of scientific success (I've heard it said that Darwinian adaptationism is the best idea that anybody's ever had, and that natural selection is the best confirmed theory in science)— *Darwinism*, of all things, undermines the scientific enterprise. Talk about biting the hand that feeds you!

There isn't, even among those who propagate this rumor, much consensus as to what one ought to do about it; indeed, there's currently a lively philosophical discussion on just that topic. Some say "well, then, so much the worse for truth." Others say "well, then, so much the worse for science." Still others say, "well, then, the paradox must be merely apparent; nothing true can refute itself."

In this discussion, I'll take a different line from any of these. My theme is that we don't need to worry about what to do if Darwinism threatens scientism because, in fact, it doesn't. There is, I claim, nothing at all in evolutionary theory that entails, or suggests, or even gives one grounds to contemplate denying the commonsense thesis that scientific inquiry quite generally leads to the discovery of objective empirical truths. I'll try to convince you of this by looking, in some detail, at the arguments that are supposed to show the contrary. I hope you won't find the process tedious; this sort of inquiry can be edifying and, anyhow, it's what philosophers mostly do for a living.

Perhaps I should mention, in passing, that I'm not actually an adaptationist about the mind; I doubt, that is, that Darwinism is true of the phylogeny of cognition. For reasons I've set out elsewhere (see the two preceding chapters) I think our kinds of minds are quite likely "hopeful monsters," which is why there are so few of them. It would thus be open to me to take the line that if scientism is jeopardized by adaptationism about the evolution of the cognitive mind, so much the worse for adaptationism about the evolution of the cognitive mind. But, though I could do that, on balance I'd rather not. That's because I'm much more confidant that scientism is true than I am that Darwinism about the phylogenesis of cognition is false. I want to leave open the possibility that they might *both* be true. I'm not, therefore, about to concede that they aren't compatible. For present purposes, then, I propose to suppress my qualms and pretend that I believe that our minds evolved; and that they did so under selection pressures. The question of interest is what, if anything, this implies about the likely truth of scientism.

So, what is the argument from adaptationism about cognition (henceforth, for short, the argument from "Darwinism") to the rejection of scientism? Actually, it's an inference that comes in several different packages. I propose to look at two. Here, to begin with, is a quotation from Patricia Churchland (1987): "There is a fatal tendency to think of the brain as essentially in the fact-finding business.... Looked at from an evolution-ary point of view, the principle function of nervous systems is to get the body parts where they should be in order that the organism may sur-vive.... Truth, whatever that is, definitely takes the hindmost."

Clearly, the crux of this quote is the claim that "the principle function of nervous systems is [not the fixation of true beliefs, but] to get the body parts where they should be in order that the organism may survive." Just why does Churchland think that "the evolutionary point of view" implies this? It's not, of course, that there's been some hot new paleobiological discovery she's just heard about ("Flash: Fossil Nervous System Found Functioning in China"). Rather, what she apparently has in mind is some inference from Darwinist general principles. Perhaps it's this: *Organisms* get selected for getting their bodies where they should be; so *nervous sys-tems* get selected for getting the bodies of organisms to where they should be; so *the function* of nervous systems—the function for which selection designed them—*is* to get the bodies of organisms to where they should be. A fortiori, the function of nervous systems isn't "fact-finding."

But if that is the intended inference, it commits a sort of distributive fallacy; it's like arguing that, since the army is numerous, each of the sol-diers must be numerous too. Consider the following parody: "Selection pressures favor reproductive success; the design of the heart was shaped by selection pressures; so the function of the heart is to mediate repro-ductive success." Parallel arguments would show that the function of *every* organ is to mediate reproductive success; hence that every organ has the same function.

Poppycock! The function of the heart—the function that its design reflects—is to circulate the blood. There is no paradox in this, and noth-ing to affront even the most orthodox Darwinist scruples. Animals that have good hearts are selected for their reproductive success as compared to animals that have less good hearts or no hearts at all. But their repro-ductive success is produced by a biological division of labor between their organs, and it's the function of the heart *relative to this division of labor* that the heart was designed to perform.

Likewise, it's entirely possible that the kind of mental architecture that maximizes behavioral adaptivity is also one that institutes a division of labor: Perhaps a cognitive mechanism that is specialized to figure out what state the world is in interacts with a conative mechanism that is specialized to figure out how to get what one wants from a world that is in that state. That's, roughly, the sort of nervous system that you'd expect if creatures act out of their beliefs, desires, and practical intelligence; and there's nothing in Darwin to suggest that they do otherwise. All Darwin adds is that, if that *is* how a creature works, then, on average, it will work better, and last longer, and have more progeny in proportion as its beliefs are true and its desires are appropriate to its ecological condition. Who would deny it?

Churchland's argument doesn't convince. But maybe there's better on offer? Here's a line that Alvin Plantinga (forthcoming; all Plantinga references are to this manuscript) has recently been pushing:[1] "Our having evolved and survived makes it likely that our cognitive faculties are reliable and our beliefs are for the most part true, only if it would be impossible or unlikely that creatures more or less like us should behave in fitness-enhancing ways but nonetheless holds mostly false beliefs" (5). But, in fact, it isn't impossible or unlikely that creatures more or less like us should behave in fitness-enhancing ways but nonetheless hold mostly false beliefs. "[T]he simplest way [of seeing how evolution might get away with selecting for cognitive mechanisms that produce generally false beliefs] is by thinking of systematic ways in which ... beliefs could be false but still adaptive.... [Imagine a creature that] thinks all the plants and animals in [its] vicinity are witches, and [whose] ways of referring to them all involve definite descriptions entailing witchhood.... [T]his would be entirely compatible with [its] beliefs' being adaptive; so it is clear ... that there would be many ways in which [its] beliefs could be for the most part false, but adaptive nonetheless." (10).

It looks like the argument that Plantinga has on offer is something like this:

> *Argument A*: Our minds evolved, so we can assume that our behavior is mostly adaptive. But our behavior could be mostly adaptive even if our beliefs were mostly false. So there's no reason to think that most of our beliefs are true.

At worst, Argument A seems better than Churchland's since it concedes, straight off, that the mind is a belief-making mechanism: Plantinga's point is that you can imagine the mind's succeeding as a belief-making mechanism even if the beliefs that it makes aren't true. Still, I don't think that Argument A could be exactly what Plantinga has in mind; not, at least, if what he has in mind is an argument against scientism (or against what he calls "naturalism," which, at least for present purposes, we may take to be much the same thing). But I do think that there's quite an interesting argument in the immediate vicinity of Argument A; one that has every right to be taken seriously. We'll come back and fix this presently. First, however, I want to grumble about a bit about a premise.

It's important to Plantinga that it not be in doubt that a system of mostly false beliefs could be adaptive. If there were anything wrong with the notion of a system of mostly false but adaptive beliefs, Plantinga's Darwinian argument against scientism couldn't get off the ground. Plantinga's way of assuring us that there's nothing wrong with this notion is to sketch a general procedure for constructing such a system: Start with our beliefs, but replace all the referring expressions by complex demonstratives (or definite descriptions) with false presuppositions. We think *that tree is very big*; they think *that witchtree is very big*; we think *the cat on the mat is asleep*; they think *the witchcat on the witchmat is asleep*. And so on. What you end up with is a creature whose behavior is plausibly about as adaptive ours, but most of whose beliefs are false.

Now I do think that goes a little fast. After all, one needs to explain *why* the behavior of a creature thus epistemically situated would be largely successful; and the intuition, surely, is that, on any way of counting them that's defensible, "most of the beliefs" that mediate its behavioral successes turn out to be true after all. Let it be that *that appletree witch is blooming* is false, or lacks a truth value, in virtue of its presupposing that that appletree is a witch. Still, much of what a creature believes in virtue of which it believes that *that appletree witch is blooming* (and in virtue of which the thought that *that appletree witch is blooming* leads to behavioral successes) are perfectly straightforwardly true. For example: *that's an appletree; that's blooming; that's there; something is blooming; something blooming is there*, and so on indefinitely. The point is trivial enough: If a creature believes *that appletree witch is blooming*, then it presumably believes that *that's an appletree* and that *that's a witch* and that *that's blooming*. And two of these are *true* beliefs that the creature shares with us, and that enter into the explanation of its behavioral successes vis-à-vis blooming appletrees in much the same way that the corresponding beliefs of ours enter into the explanation of our behavioral success vis-à-vis blooming appletrees. Notice that this argument doesn't depend on assuming (as, in fact, I don't) that belief is closed under implication, or even under "immediate" implication. All it needs is the truism that the creature has whatever beliefs the explanations of its behavioral successes require attributing to it. This, surely, is common ground *whether or not* the beliefs that such explanations appeal to are supposed to be true.

The philosopher Berkeley seems to have believed that tables and chairs are logically homogeneous with afterimages. I assume that he was wrong to believe this. But intuition strongly suggests that the truth of "most of" his beliefs survived his metaphysical aberrations, and that it was these residual true beliefs that explain his behavioral successes. He may, for example, have chosen to sit in this chair rather than that one because he believed that this afterimage chair is more comfortable than that afterimage chair. And that may have been the right choice in the circumstance. If so, we should credit Berkeley's success to his true belief that this chair is comfortable, and not to his false belief that this chair is an afterimage.

The long and short is: There's admittedly no agreed-on way of counting beliefs; but it's not clear that the beliefs of the creature Plantinga asks us to imagine are mostly false. Or, indeed, that its epistemic situation actually differs very much from our own. So, suppose that the forces of natural selection can't distinguish that creature from us; suppose, that is to say, that its aggregate beliefs are about as adaptive as ours. Even so, so far we don't have a crystal clear case of the kind that Plantinga needs: one where selection might favor a creature whose beliefs are, in some radical sense, "mostly false."

If I'm right about this, it leaves us at a bit of an impasse. I'm pretty sure that Plantinga's way of constructing a system of beliefs that are clearly mostly adaptive but mostly false doesn't work. On the other hand, I haven't an argument for what I strongly suspect, that there is *no* way of constructing such a system. And I don't want the discussion to break down here, because I think the following question is worth considering: *If it is possible to construct a system of adaptive but mostly false beliefs, is there then an argument from Darwinism to the rejection of scientism?*

So here's what I propose. Let's just assume that there is a "charmed circle" of nonstandard cognitive mechanisms such that:

- every cognitive mechanism in the charmed circle produces systems of mostly false beliefs, and
- most creatures that act on such systems of false beliefs (most "false believers" as I'll say) will do *at least as well* as most creatures that act on the corresponding true beliefs (i.e., at least as well as most corresponding "true believers").

It is very nice of me to concede all this since I really am much inclined to doubt that it's true.

We need to do one more preliminary thing before we can get to the serious stuff. As we've seen, Plantinga writes that "Our having evolved and survived makes it likely that our cognitive faculties are reliable and our beliefs are for the most part true, only if it would be impossible or unlikely that creatures more or less like us should behave in fitness-enhancing ways but nonetheless hold mostly false beliefs" (5).

Now, that's okay as it's stated because, as it's stated, it only purports to give a *necessary* condition for

(1)   Our having evolved makes it likely that our cognitive faculties are reliable.

But what Plantinga actually needs, if he's to have an argument against scientism, is something like a *sufficient* condition for

(2)   Our having evolved makes it likely that our cognitive faculties are unreliable.

Short of arguing for (2), the most Plantinga's got is that Darwinism is *compatible* with scientism's being false; which I suppose to be true but not interesting. That the cat is on the mat is *also* compatible with scientism's being false. So what? The cat's being on the mat isn't therefore *a reason to think* that scientism *is* false. Churchland appears to make a similar mistake: If evolution selected our minds for moving our bodies, then perhaps Darwinism gives us no reason to suppose that our beliefs are mostly true. But that's quite different from Darwinism giving us some reason to suppose that our beliefs are mostly *not* true, and it's this latter than the Darwinian argument against scientism requires.

In short, what Plantinga really needs (and what I haven't any doubt is what he really intends) is argument A*. Argument A* is just like Argument A except that the conclusion is stronger.

> *Argument A**: Our minds evolved, so we can assume that our behavior is mostly adaptive. But our behavior could be mostly adaptive even if most of our beliefs were mostly false. So there is reason to think that most of our beliefs *are* false.

Now, if I were in a bad mood, or if I were generally an unobliging chap, I could, with some justice, harp on the difference between arguments A and A*. Consider:

> *Argument B*: I am standing here giving a lecture and I am wearing my gray socks. But there are other colors (green, say) such that, quite likely, someone a lot like me might be standing here giving a lecture if he were wearing socks of any of those colors. So *that I am wearing gray socks* is not a reason to believe *that I am standing here giving a lecture*.

Argument B seems alright; that I am wearing gray socks *isn't* a reason to believe that I am standing here giving a lecture. It's just that, as a matter of fact, both happen to be true. But, notice, argument B* doesn't seem right at all.

> *Argument B**: I am standing here giving a lecture and I am wearing gray socks. But there are other colors of socks (green, say) such that, quite likely, someone a lot like me might be standing here giving a lecture if he were wearing socks of any of those colors. *So that I am wearing gray socks is a reason to believe that I am not standing here giving a lecture*.

Poppycock once again. But, as far as I can see, the difference between B and B* is just like the difference between A and A*, and it's important not to equivocate between A and A* for just the same reason that it is important not to equivocate between B and B*. I'm a little worried that Plantinga's Darwinian argument against scientism trades on this sort of equivocation; I haven't, anyhow, found a way to make it run that clearly doesn't.

So, as I say, if I were in a bad mood or generally unobliging I could make a fuss of this. But I'm not going to because, as it turns out, my main line of objection to Plantinga is good against both A* and A. Nothing that follows, therefore, will depend on the difference between them (except at the very end).

Recapitulation: I've agreed to assume that the charmed circle is non-empty. By this assumption, it's likely that there could be false believers that would, in principle, be just as successful as true believers are. What I need to show is that, pace Plantinga, this assumption doesn't even come *close* to underwriting either A or A*.

Well, it doesn't. Here's why: What I've just conceded is, in effect, that what evolution "really wants" is not true believers, but *either* true believers *or* just any old believers whose cognitive mechanisms are in the charmed circle. But the question nonetheless arises, *how, in our case, did evolution go about getting what it wanted?* Darwinism is, remember, a *historical* thesis about *us*. If it's to jeopardize scientism, it's got to underwrite the claim that our minds actually evolved under selection that favored us *because* our false beliefs were generally successful. That there are, in principle, systems of generally successful false beliefs isn't enough to show this; nor is it enough that there quite possibly could be creatures a lot like us whose minds evolve under selection for false beliefs and whose behaviors are generally successful. What's required, to repeat, is a plausible historical scenario according to which *our* minds were selected because we had lots of behaviorally successful false beliefs. Well, is there such a scenario? The answer depends on the status of the following claim:

> *There is some property P of our beliefs such that:*
> i. In the past, our acting on false beliefs that have P actually has been (at least) as successful as our acting on true beliefs was. (Otherwise selection would have had a reason for preferring true beliefs to false beliefs with P.)
> ii. It's reliable (counterfactual supporting) that false beliefs with P are at least as behaviorally successful as true beliefs. (That is, at least some cognitive mechanisms that produce P beliefs are in the charmed circle.)
> iii. Assuming our cognitive system is still doing what it evolved to do—that is, still performing the function for which it was selected—our current behavioral successes are *still* largely predicated on false beliefs that have P.

The reason there has to be such a property is simply that, according to the present story, evolution selected us *because of* something about lots of our false beliefs that reliably lead to our behavioral success. So the present story couldn't be true unless there *was* something about lots of our false beliefs that reliably lead to our behavioral successes.

Well, was there? I doubt it, and (short of philosophical skepticism), nobody I've heard of actually believes it. What we all believe is that when actions out of false beliefs are successful, that's generally a *lucky accident*; and, correspondingly, that a policy of acting on false beliefs, even when it works in the short run, generally gets you into trouble sooner or later. Indeed—and this is the point I want to stress—the commonsense view is that we have, even as we stand here now, lots of *solid inductive evidence* that our false beliefs pretty generally don't have property P. The inductive evidence is that, though some of our actions on false beliefs have succeeded from time to time, pretty generally our actions on false beliefs have *failed*; and they've failed precisely *because* they were actions on false beliefs. Conversely, in what must surely be overwhelmingly the standard case, if the explanation of a behavioral success invokes a creature's beliefs and desires at all, it does so as follows: The creature acted in such and such a way because it wanted to bring it about that so and so; and because it believed, truly, that acting that way *would* bring it about that so and so. That much about intentional explanation has been clear since Aristotle. Perhaps, when you're doing philosophy, you're prepared to doubt it; but you don't, surely, doubt it when you're being serious.

This commonsense view of the relation between true belief and successful behavior may be wrong, of course, but I want to stress that you can't invoke the *assumption* that it's wrong on behalf of the putative Darwinian argument against scientism. For, that we are probably false believers is supposed to be the *conclusion* of the Darwinian argument, not its *premise*. It is very easy when one is doing epistemology—especially when one is doing skeptical epistemology—to find oneself arguing backwards. I want to pause to rub this point in a bit, because I think it may really be at the bottom of a lot of the trouble.

You clearly could get the required antiscientistic Darwinian conclusion from the following argument *assuming that its premises are true*.

> *Argument C*: 1. Many (/most) of our behavioral successes probably are (/probably have been) predicated on false beliefs that have P.
> 2. So selection has had inductive grounds for favoring false beliefs that have P. 3. So we've probably evolved to have false beliefs that have P.

But, surely, Argument C is question begging in the present context. As it stands it's no better than the first premise, and on the face of it, the first premise is not plausible. What's needed to make the argument convincing is some *independent* reason for thinking that many of our behavioral successes actually have been predicated on false beliefs that have *P*. And, on pain of circularity, it better be a reason that doesn't itself assume that we've probably evolved to have false beliefs. Short of philosophical skepticism, I know of no independent reason to believe premise 1 and, as remarked above, all the inductive evidence I've heard of is to the contrary.

I hope the main line of my argument against Plantinga is now getting clear. It goes in three steps.

Step 1: Believing that there is no such property as *P* is *not at all* the same as believing that the charmed circle is empty, or believing that it's *likely* that the charmed circle is empty. I, for example, believe that, in point of historical fact, those of my successful actions that have been predicated on false beliefs were mostly flukes; a fortiori, I believe that they weren't predicated on false beliefs that had *P*. But I also believe there could be cognitive systems of the sort that membership in the charmed circle requires. All *that* needs is some property such that *if* my false beliefs had had it, my actions *would have been* reliably successful. It's quite consistent of me to believe that there very likely *are* properties such that if my false beliefs had had them they would have been reliably successful, but to doubt that the false beliefs on which my successful actions have *actually* been predicated actually *have* had any of these properties. And likewise for the false beliefs on which the successful actions of my grandmother were predicated; and likewise for those of her grandmother; and likewise back to the protoplasmic slime.

Step 2: For it to be true that I was selected because in the past my false beliefs had *P*, it would at least have to be true that a lot of my past behavioral successes have been predicated on false beliefs tout court. But, in fact, I have no reason at all to believe that lots of my past behavioral success have been predicated on false beliefs, and I have vast lots of inductive evidence to the contrary. For example, this morning I managed to get my teeth brushed; a small behavioral success, to be sure, but mine own. Such as it was, I'm quite certain that it was prompted and guided by a host of true beliefs including, inter alia: true beliefs about my teeth needing a brush; true beliefs about the spatiotemporal location of my toothbrush; true beliefs about the spatiotemporal location of my teeth; true beliefs about the spatiotemporal location of my limbs with respect to my tooth brush and my teeth, and so on. Certainly, short of philosophical skepticism, I can think of no reason in the world to deny any of this.

I pause to emphasize the caveat. I said that I can think of no reason *short of philosophical skepticism* for doubting the commonsense view that most of one's successful actions are and have been actions out of true beliefs. This way of putting it is *not* question begging in the present context. For, I am not trying to refute philosophical skepticism, nor do I doubt that Darwinism *together with* philosophical skepticism would argue that we probably aren't true believers. That must be so since philosophical skepticism, all by itself, would argue that we probably aren't true believers if only *it* were true. None of that is pertinent, however, in the present context; the question before the house is whether Darwinism undermines scientism from, as it were, the inside; namely, *without* supplementary philosophical commitments.

Step 3: If the mere nonemptiness of the charmed circle proves nothing, and if the available inductive evidence is that our successful actions are quite generally predicated on true beliefs, then all that's left for the Darwinian argument against scientism is to suppose that things have changed a lot since our minds got selected: Whatever the situation is *now*, it *used to be* that our successful behavior was generally predicated (not on true beliefs but) on false beliefs that have P. That's how it was back when selection molded cognition.

But why on earth should anyone believe this? And, even if anyone does believe it, why does it matter in the present context? That minds used to achieve their successes by causing us to act on false beliefs is no challenge to scientism *as long as they don't do it that way any more*. The relevant consideration is one that Gould and Lewontin stressed in their famous paper about spandrels: "[there are cases where one finds] adaptation and selection but the adaptation is a secondary utilization of parts present for reasons of architecture, development, or history. If blushing turns out to be an adaptation affected by sexual selection in humans, it will not help us to understand why blood is red" (1979, 159). Correspondingly, even if it turns out that nervous systems were originally selected for making adaptive false beliefs—a hypothesis which I don't believe for a minute, and which we've thus far been offered no reason to grant—that *still* wouldn't ground a Darwinian argument against scientism.

Now we're ready to recap. The polemical situation, according to the opposition, is that a Darwinist has no reason for supposing that evolution picks true believers rather than false-believers-with-cognitive-mechanisms-in-the-charmed-circle. This is because, by assumption, the behavior of the latter would be just as adaptive as the behavior of the former. But, I claim, that's not good enough. It can't be plausible that we were selected for the behavioral success of our false beliefs unless it is *independently* plausible (or, anyhow, independently not *im*plausible) that a lot of our behavioral successes have in fact been actions out of false beliefs. But, to repeat, the inductively plausible commonsense view is that, however we may have evolved, our behavioral successes are generally predicated on *true* beliefs; and that it's generally the truth of the beliefs on which they are predicated that explains the successes of our behaviors;

and that, in cases when we have managed to act successfully out of false beliefs, that's mostly been blind luck. As far as I can tell, there is no reason on God's green earth why a proper Darwinism should reject this commonsense view.

I propose to conclude with a caveat. It's been one of my complaints against Churchland and Plantinga that they seem to be inclined to argue from "Darwinism doesn't imply that most of our beliefs are true," to "it's therefore likely that most of our beliefs are false"; an argument which is, as I remarked, invalid on the face of it. But I also want to warn against the converse fallacy, that of arguing from "Darwinism doesn't imply that most of our beliefs are false" to "it's therefore likely that most of our beliefs are true."

A fair lot of philosophers have been tempted by the idea that, far from refuting scientism, the theory of evolution actually underwrites it. Thus Dan Dennett writes (1978a, 17), "[T]he capacity to believe would have no survival value unless it were a capacity to believe truths." What he's endorsing is not the casual truism that any creature that is able to believe falsely that P is likewise able to believe truly that not-P; it's the much stronger claim that the Darwinian survival of a thinking creature ipso facto implies, or and how makes plausible, the truth of most of what that creature believes.

But that simply isn't so, and the reason is much the one that Plantinga suggests. That a creature evolved to have some organ is, if the Darwinian story is true, something like a sufficient argument that having that organ was adaptive for the creature. But it is not part of the Darwinian story, nor is it independently the case, that if a creature *failed* to evolve to have some organ, that's anything like a sufficient argument that having it *wouldn't have been* adaptive for the creature. Darwinism deals in *sufficient* conditions for adaptivity, not in *necessary* conditions for adaptivity.[2] The point this observation turns on is, in fact, perfectly general: Darwinism isn't about which traits of organisms *would have been adaptive if a creature had had them*. Darwinism is about what effects the actual possession of a trait has as a function of its adapativity: roughly, the frequency with which a trait is represented in a population is proportional to its adaptivity, all else equal.

To put much the same point the other way round: What Darwinism requires as (roughly) sufficient for the evolution of a trait is *both* that the trait would be adaptive if the organism were to have it; *and* that, in point of historical fact, there are some (rudimentary, approximate, proto-) instances of the trait actually on the ground for selection to operate upon. I've supposed that what was on the ground when our minds evolved were primitive cognitive mechanisms that produced, by and large, adaptive true beliefs; and that selection favored, among such cognitive mechanisms, the ones that produced adaptive true beliefs efficiently and reliably and about a wide range of ecologically salient states of affairs. The reason I've supposed this is that, pretty clearly, it's mostly true beliefs that eventuate in adaptive behavior *now*. But, to repeat, Darwin would have been just as happy if what had been on offer for selection to choose from when we evolved was cognitive mechanisms which produce, by and large, adaptive *false* beliefs; in that case, the Darwinian prediction would have been that it's mostly false beliefs that eventuate in adaptive behavior now. It is, just as Plantinga and Churchland both say, the adaptivity of the beliefs, and not their truth per se, that evolution cares about. Since, for all we know, evolution would have chosen false believers if it had been given the chance, the fact that evolution chose *us* isn't, in and of itself, a reason for thinking that we're true believers. There is, as far as I can tell, *no* Darwinian reason for thinking that we're true believers. Or that we aren't.

The problems of philosophy are very hard, so we keep hoping that somebody else will take them off our hands. Darwin, Heaven help him, has been for years among our favorite candidates. In the nineteenth century, we thought that maybe he would do our ethics for us. These days, it's the burden of grounding our epistemology that we want him to carry. According to temperament, we'd either like him to show that most of our beliefs have to be true or that most of them have not to be. Either way, we want Darwin to demarcate, once and for all, the necessary limits of human fallibility. He's done so much for biologists; why can't he do this one little thing for philosophers?

But Darwin isn't in the epistemology business, and evolution *doesn't care* whether most of our beliefs are true. Evolution is *neutral* as to whether most of our beliefs are true. Like Rhett Butler in the movie, it just doesn't give a damn.

So, then, why do I believe in scientism? Or, to put the question slightly differently: If the theory of natural selection doesn't imply that our empirical beliefs are generally pretty close to being true, then why, in particular, do I believe that the lunar theory of the tides, and the tectonic theory of the arrangement of the continents, are pretty close to being true? Put that way, the question sort of answers itself. I believe these theories on account of my (casual, to be sure) acquaintance with the evidence that has been alleged for them. And, of course, the evidence that's been alleged for the lunar theory of the tides isn't that human cognition evolved; it's that, for example, the lunar theory correctly predicts that (all else equal) the tides are highest when the moon is full. Likewise, the evidence alleged for the tectonic theory of the arrangement of the continents isn't that human cognition evolved; it's that, for example, the geology of Western Africa matches the geology of Eastern South America in ways that are otherwise quite inexplicable. This is just what you'd expect; and it's just what the scientists themselves take for granted. What on earth does whether or not our minds evolved have to do with whether or not it's the moon that pulls the tides around?

Maybe, though I very much doubt it, the whole scientific enterprise somehow requires, or presupposes, an epistemological foundation to rest on. And if philosophers can't get it one a priori by an exercise of Cartesian intuition, maybe they can get it one some other way. Well, maybe. But Darwin isn't going to help. On the contrary, since the theory of evolution is itself a part of the scientific enterprise, whatever epistemological boat we're in, Darwin's in there with us. My scientism is not in the least disturbed by this. Quite the opposite, it's perfectly delighted to have him aboard.

Acknowledgment

I'm grateful to Professor Plantinga for some email exchanges about this material, and for allowing me to read Plantinga (forthcoming) in a manuscript version. Plantinga doesn't think I've got to the bottom of the arguments he gives there. No doubt he's right.

Notes

1. There is, in fact, more to Plantinga's argument than what I will discuss here. Plantinga thinks (as I do not) that there is some serious chance it will turn out that we don't act out of the content of our mental states *at all*, and a fortiori that the truth of our beliefs can't be what explains our behavioral successes. In any event, I'll consider only the aspects of Plantinga's polemic that turn on Darwinst considerations.
2. This is quite a general property of teleological explanations, as, by the way, Carl Hempel pointed out some while ago (Hempel, 1965).

Chapter 17

Review of Steven Pinker's *How the Mind Works* and Henry Plotkin's *Evolution in Mind*

It belongs to the millennial mood to want to sum things up, see where we have gotten, and point the direction where further progress lies. Cognitive science has not been spared this impulse, so here are two books purporting to limn the state of the art. They differ a bit in their intended audience; Plotkin's is more or less a text, while Pinker hopes for a general readership. Pinker covers much more ground, but he takes an ungainly 600 pages to do it, compared to Plotkin's svelte 275. Both authors are unusually good at exposition, Pinker exceptionally so from time to time. And their general sense of what's going on and of what should come next are remarkably similar considering that they are writing about a field that is notoriously fractious. Taken severally or together, they present what is probably the best statement you can find in print of a very important contemporary view of mental structure and process.

But how much of this view is true? To begin with, Pinker and Plotkin are reporting a minority consensus. Most cognitive scientists still work in a tradition of empiricism and associationism, whose main tenets haven't changed much since Locke and Hume. The human mind is a blank slate at birth. Experience writes on the slate, and association extracts and extrapolates whatever trends there are in the record that experience leaves. The structure of the mind is thus an image, made a posteriori, of the statistical regularities in the world in which it finds itself. I would guess that quite a substantial majority of cognitive scientists believe something of this sort; so deeply, indeed, that many hardly notice that they believe it.

Pinker and Plotkin, by contrast, epitomize a rationalist revival that started about forty years ago with Chomsky's work on the syntax of natural languages, and that is by now sufficiently robust to offer a serious alternative to the empiricist tradition. Like Pinker and Plotkin, I think the new rationalism is the best story about the mind that science has found to tell so far. But I think their version of that story is tendentious; indeed importantly flawed. And I think the cheerful tone that they tell it in is quite unwarranted by the amount of progress that has actually been made. Our best scientific story about the mind is better than empiricism; but, in all sorts of ways, it's still not very good. Pinker quotes Chomsky's remark that "ignorance can be divided into problems and mysteries.... I wrote this book [Pinker continues] because dozens of mysteries of the mind, from mental images to romantic love, have recently been upgraded to problems (though there are still some mysteries too!)" (ix) Well, cheerfulness sells books, but what Ecclesiastes said still holds: "The heart of the wise is in the house of mourning."

Pinker elaborates his version of rationalism around four main ideas:

- The mind is a computational system.
- The mind is massively modular.
- A lot of mental structure, including a lot of cognitive structure, is innate.
- A lot of mental structure, including a lot of cognitive structure, is an evolutionary adaptation. In particular, the function of a creature's nervous system is to abet the propagation of its genome (its selfish genes, as one says).

Plotkin agrees with all four of these theses, though he puts less emphasis than Pinker does on the minds-are-computers part of the story. Both authors take for granted that psychology should be a part of biology, and they are both emphatic about the need for more Darwinian thinking in cognitive science. (Plotkin quotes with approval Theodore Dobzhansky's dictum that "nothing in biology makes sense except in the light of evolution," amending it, however, to read "makes complete sense.") It's their Darwinism, specifically their allegiance to a "selfish gene" account of the phylogeny of mind, that most strikingly distinguishes Pinker and Plotkin from a number of their rationalist colleagues (from Chomsky in particular). All this needs some looking into. I'll offer a sketch of how the four pieces of the Pinker-Plotkin version of rationalism are connected; and, by implication, of what an alternative rationalism might look like. I'm particularly interested in how much of the Pinker-Plotkin consensus turns on the stuff about selfish genes, of which I don't, in fact, believe a word.

Computation

Beyond any doubt, the most important thing that has happened in cognitive science was Turing's invention of the notion of mechanical rationality. Here's a quick, very informal, introduction. (Pinker provides one that's more extensive.)

It's a remarkable fact that you can tell, just by looking at it, that any sentence of the syntactic form *P* and *Q* ("John swims and Mary drinks," as it might be) is true if and only if *P* and *Q* are both true. "You can tell just by looking" means: to see that the entailments hold, you don't have to know anything about what either *P* or *Q* means, and you don't have to know anything about the nonlinguistic world. This really is remarkable since, after all, it's what they mean, together with how the nonlinguistic world is, that decide whether either P or Q themselves are true. This line of thought is often summarized by saying that some inferences are rational in virtue of the syntax of the sentences that enter into them; metaphorically, in virtue of the "shapes" of these sentences.

Turing noticed that, wherever an inference is formal in this sense, a machine can be made to execute the inference. This is because although machines are awful at figuring out what things mean and are not much better at figuring out what's going on in the world, you can make them so that they are quite good at detecting and responding to syntactic relations between sentences. Given an argument that depends just on the syntax of the sentences that it is couched in, such a device will accept the argument if and only if it is valid. To that extent, you can build a rational machine. Thus, in chrysalis, the computer and all its works.

Thus too the idea that some, at least, of what makes minds rational is their ability to perform computations on thoughts; where thoughts, like sentences, are assumed to be syntactically structured, and where "computations" means formal operations in the manner of Turing. It's this theory that Pinker has in mind when he claims that "thinking is a kind of computation" (21). It has proved to be a simply terrific idea. Like Truth, Beauty, and Virtue, Rationality is a normative notion; the computational theory of mind is the first time in all of intellectual history that a science has been made out of one of those. If God were to stop the show now and ask us what we've discovered about how we think, Turing's theory of computation is by far the best thing that we could offer. But.

Turing's account of computation is, in a couple of senses, local. It doesn't look past the form of sentences to their meanings; and it assumes that the role of thoughts in a mental process is determined entirely by their internal (syntactic) structure. And there's reason to believe that at least some rational processes are not local in either of these respects. It may be that wherever either semantic or global features of mental processes begin to make their presence felt, you reach the limits of what Turing's kind of computational rationality is able to explain. As things stand, what's beyond these limits is not a problem but a mystery.

For example, I think it's likely that a lot of rational belief formation turns on what philosophers call "inferences to the best explanation." You're given what perception presents to you as currently the fact, and you're given what memory presents to you as the beliefs that you've formed till now. Your cognitive problem is to find and adopt whatever new beliefs are best confirmed on balance. "Best confirmed beliefs on balance" means something like: the strongest and simplest relevant beliefs that are consistent with as many of one's prior epistemic commitments as possible. But, as far as anybody knows, relevance, strength, simplicity, centrality, and the like are properties, not of single sentences, but of whole belief systems; and there's no reason at all to suppose that such global properties of belief systems are syntactic.

In my view, the cognitive science that we've got so far has hardly begun to face this issue. Most practitioners (Pinker and Plotkin included, as far as I can tell) hope that it will resolve itself into lots of small, local problems which will in turn succumb to Turing's kind of treatment. Well, maybe; it's certainly worth the effort of continuing to try. But I'm impressed by this consideration: Our best cognitive science is the psychology of language and the psychology of perception, and (see just below) it may well be that linguistic and perceptual processes are largely modular, hence computationally local. Howerer, plausibly, the globality of cognition shows up clearest in common sense reasoning. Uncoincidentally, as things now stand, we don't have a theory of the psychology of common sense reasoning that would survive serious scrutiny by an intelligent five-year-old. Likewise, common sense is what the computers that we know how to build egregiously don't have. I think it's likely that we are running into the limits of what can be explained with Turing's kind of computation; and I think we don't have any idea what to do about it.

Suffice it, anyhow, that if your notion of computation is exclusively local, like Turing's, then your notion of mental architecture had best be massively modular. That brings us to the second tenet of the Pinker-Plotkin version of rationalism.

Massive modularity   A module is a special purpose, more or less autonomous computational system. It's built to solve a very restricted class of problems, and the information it can use to solve them with is proprietary. Most of the new rationalists think that at least some human cognitive mechanisms are modular, aspects of language and perception being among the classical best candidates. For example, the computations that convert a two-dimensional array of retinal stimulations into a stable image of a three-dimensional visual world are supposed to be largely autonomous with respect to the rest of one's cognition. That's why many visual illusions don't go away even if you know that they are illusory. Massimo Piatelli, reviewing Plotkin's book in *Nature*, remarks that the modularity of cognitive processes is "arguably ... the single most important discovery of cognitive science." At a minimum, it's what most distinguishes our current cognitive science from its immediate precursor, the "New Look" psychology of the 1950s that emphasized the continuity of perception with cognition, and hence the impact of what one believes on what one sees.

Both Pinker and Plotkin think the mind is mostly made of modules; that's the massive modularity thesis. I want to stress how well it fits the idea that mental computation is local. By definition, modular problem solving works with less than everything that a creature knows. It thereby minimizes the global cognitive effects that are the bane of Turing's kind of computation. If the mind is massively modular, then maybe the notion of computation that Turing gave us is, after all, the only one that cognitive science needs. It would be nice to be able to believe that; Pinker and Plotkin certainly try very hard to.

But it seems to me one can't. Eventually the mind has to integrate the results of all those modular computations, and I don't see how there could be a module for doing *that*. The moon looks bigger when it's on the horizon; but I know perfectly well it's not. My visual perception module gets fooled, but I don't. The question is: who is this I? And by what—presumably global—computational process does it use what I know about the astronomical facts to correct the misleading appearances that my visual perception module insists on computing? If, in short, there is a community of computers living in my head, there had also better be somebody who is in charge; and, by God, it had better be me. The old rationalists, like Kant, thought that the integration of information is a lot of what's required to turn it into knowledge. If that's right, then a cognitive science that hasn't faced the integration problem has scarcely gotten off the ground. Probably, modular computation doesn't explain how minds are rational; it's just a sort of precursor. It's what you have to work through to get a view of how horribly hard our rationality is to understand.

Innateness   Rationalists are nativists by definition; and nativism is where cognitive science touches the real world. As both Pinker and Plotkin rightly emphasize, the standard view in current social science—and in what's called "literary theory"—takes a form of empiricism for granted: human nature is arbitrarily plastic, and minds are social constructs. By contrast, the evidence from cognitive science is that a lot of what's in the modules seems to be there innately. Pinker and Plotkin both review a fair sample of this evidence, including some of the lovely experimental work on infant cognition that psychologists have done in the last couple of decades. There is also, as the linguists have been claiming for years, a lot of indirect evidence that points to much the same conclusion: All human languages appear to be structurally similar in profound and surprising ways. There may be an alternative to the nativist explanation that linguistic structure is genetically specified; but, if there is, nobody thus far has had a glimpse of it.

Cultural relativism is widely held to be politically correct. If so, then sooner or later political correctness and cognitive science are going to collide. Many tears will be shed and many hands will be wrung in public. So be it. If there is a human nature, and if it is to some interesting extent genetically determined, it is folly for humanists to ignore it. We're animals whatever else we are; and what makes an animal well, and happy, and sane depends a lot on what kind of animal it is. Pinker and Plotkin are both very good on this; I commend them to you. But, for present purposes, I want to examine a different aspect of their rationalism.

Psychological Darwinism   Pinker and Plotkin both believe that if nativism is the right story about cognition it follows that much of our psychology must be, in the Darwinian sense, an evolutionary adaptation; that is, it must be intelligible in light of evolutionary selection pressures that shaped it. It's the nativism that makes cognitive science politically interesting. But it's the inference from nativism to Darwinism that is currently divisive within the rationalist community. Pinker and Plotkin are selling an evolutionary approach to psychology that a lot of cognitive scientists (myself included) aren't buying. I want to spend some time on this.

There are two standard arguments, both of which Pinker and Plotkin endorse, that are supposed to underwrite the inference from nativism to psychological Darwinism. The first is empirical, the second is methodological. I suspect that both are wrong headed.

The empirical argument is that, as a matter of fact, there is no way except evolutionary selection for nature to build a complex, adaptive system. Plotkin says that "neo-Darwinian theory [is] the central theorem of all biology, including behavioral biology," (54) and that "If behavior is adaptive, then it must be the product of evolution" (53). Likewise Pinker: "Natural selection is the only explanation we have of how complex life can evolve ... [so] natural selection is indispensable to understanding the human mind" (155).

One reply to this argument is that there is, after all, an alternative to natural selection as the source of adaptive complexity; you could get some by a miracle. But I'm not a creationist, nor are any of my rationalist friends, as far as I know. Nor do we have to be, since there's another way out of the complexity argument. This is a long story, but here's the gist: It's common ground that the evolution of our behavior was mediated by the evolution of our brains. So, what matters to the question of whether the mind is an adaptation is not how complex our behavior is, but how much change you would have to make in an ape's brain to produce the cognitive structure of a human mind. And about this, exactly nothing is known. That's because nothing is known about how the structure of our minds depends on the structure of our brains. Nobody even knows *which* brain structures our cognitive capacities depend on.

Unlike our minds, our brains are, by any gross measure, very similar to those of apes. So it looks as though relatively small alterations of brain structure must have produced very large behavioral discontinuities in the transition from the ancestral apes to us. If that's right, then you don't have to assume that cognitive complexity is shaped by the gradual action of Darwinian selection on prehuman behavioral phenotypes. Analogies to the evolution of organic structures, though they pervade the literature of psychological Darwinism, actually don't cut much ice here. Make the giraffe's neck just a little longer and you correspondingly increase, by just a little, the animal's capacity to reach the fruit at the top of the tree. So it's plausible, to that extent, that selection stretched giraffes' necks bit by bit. But make an ape's brain just a little bigger (or denser, or more folded, or, who knows, greyer) and it's anybody's guess what happens to the creature's behavioral repertoire. Maybe the ape turns into us.

Adaptationists about the phylogeny of cognition say that it's a choice between Darwin and God, and they like to parade as scientifically tough-minded about which one of these you should pick. But that misstates the alternatives, so don't let yourself be bullied. In fact, we don't know what the scientifically reasonable view of the phylogeny of behavior is; nor will we until we begin to understand how behavior is subserved by the brain. And bother tough-minded anyhow; what matters is what's true.

The methodological argument from nativism to Darwinism fares no better. Pinker and Plotkin both hold that the (anyhow, a) proper method of cognitive psychology is "reverse engineering." Reverse engineering is inferring how a device must work from, inter alia, a prior appreciation of its function. If you don't know what a can opener is for, you are going to have trouble figuring out what its parts do. In the case of more complex machines, like for example people, your chance of getting the structure right is effectively nil if you don't know the function. Psychological Darwinism, so the argument goes, gives us the very notion of function that the cognitive scientist's reverse engineering of the mind requires: To a first approximation, and with, to be sure, occasional exceptions, the function of a cognitive mechanism is whatever evolution selected it for. Without this evolutionary slant on function, cognitive science is therefore simply in the dark. So the argument goes.

This too is a long story. But if evolution really does underwrite a notion of function, it is a historical notion; and it's far from clear that a historical notion of function is what reverse engineering actually needs. You might think, after all, that what matters in understanding the mind is what ours do now, not what our ancestors' did some millions of years ago. And, anyhow, the reverse engineering argument is over its head in anachronism. As a matter of fact, lots of physiology got worked out long before there was a theory of evolution. That's because you don't have to know how hands (or hearts, or eyes, or livers) evolved to make a pretty shrewd guess about what they are for. Maybe you likewise don't have to know how the mind evolved to make a pretty shrewd guess at what it's for; for example, that it's to think with. No doubt, arriving at a "complete" explanation of the mind by reverse engineering might require an appreciation of its evolutionary history. But I don't think we should be worrying much about complete explanations at this stage. I, for one, would settle for the merest glimpse of what is going on.

One last point about the status of the inference from nativism to psychological Darwinism. If the mind is mostly a collection of innate modules, then pretty clearly it must have evolved gradually, under selection pressure. That's because, as I remarked above, modules contain lots of specialized information about the problem domains that they compute in. And it really would be a miracle if all those details got into brains via a relatively small, fortuitous alteration to some ape's neurology. To put it the other way around, if adaptationism isn't true in psychology, it must be that what makes our minds so clever is something pretty general; something about their global structure. The moral is that if you aren't into psychological Darwinism, you shouldn't be into massive modularity either. Everything connects.

For the sake of the argument, however, let's suppose that the mind is an adaptation after all and see where that assumption leads. It's a point of definition that adaptations have to be for something. Pinker and Plotkin both accept the "selfish gene" story about what biological adaptations are for. Organic structure is (mostly) in aid of the propagation of the genes. And so is brain structure inter alia. And so is cognitive structure, since how the mind works depends on how the brain does. So there's a route from Darwinism to sociobiology; and Pinker, at least, is keen to take it. (Plotkin seems a bit less so. He's prepared to settle for arguing that some of the apparently egregious problems for the selfish gene theory—the phylogeny of altruism, for example—may be less decisive than were at first supposed. I think that settling for that is very wise of him.)

A lot of the fun of Pinker's book is his attempt to deduce human psychology from the assumption that our minds are adaptations for transmitting our genes. His last several chapters are devoted to this, and they range very broadly, including, so help me, one on the meaning of life. Pinker would like to convince us that the predictions that the selfish gene theory makes about how our minds must be organized are independently plausible. But this project doesn't fare well either. Prima facie, the picture of the mind, indeed of human nature in general, that psychological Darwinism suggests is preposterous; a sort of jumped up, down market version of original sin.

Psychological Darwinism is a kind of conspiracy theory; that is, it explains behavior by imputing an interest (viz., one in the proliferation of the genome) that the agent of the behavior does not acknowledge. When literal conspiracies are alleged, duplicity is generally part of the charge: "He wasn't making confetti; he was shredding the evidence. He did X in aid of Y, and then he lied about his motive." But in the kind of conspiracy theories that psychologists like best, the motive is supposed to be inaccessible even to the agent, who is thus perfectly sincere in denying the imputation. In the extreme case, it hardly even is the agent to whom the motive is attributed. Freudian explanations provide a familiar example: What seemed to be merely Jones's slip of the tongue was the unconscious expression of a libidinous impulse. But not Jones's libidinous impulse, really; one that his Id had on his behalf. Likewise, for the psychological Darwinist: What seemed to be your (after all unsurprising) interest in your child's well-being turns out to be your genes' conspiracy to propagate themselves. Not your conspiracy, but theirs.

How do you make out the case that Jones did X in aid of an interest in Y, when Y is an interest that Jones doesn't own up to? The idea is perfectly familiar: You argue that X would have been the rational (reasonable, intelligible) thing for Jones to do if Y had been his motive. Such arguments can be very persuasive. The files Jones shredded were precisely the ones that incriminated him; and he shredded them in the middle of the night. What better explanation than that Jones conspired to destroy the evidence? Likewise when the conspiracy is unconscious. Suppose that an interest in the propagation of the genome would rationalize monogamous families in animals whose offspring mature slowly. Well, our offspring do mature slowly; and our species does, by and large, favor monogamous families. So that's evidence that we favor monogamous families *because* we have an interest in the propagation of our genes. Well, isn't it?

Maybe yes, maybe no; this kind of inference needs to be handled with great care. For, often enough, where an interest in X would rationalize Y, so too would an interest in P, Q, or R. It's reasonable of Jones to carry an umbrella if it's raining and he wants to keep dry. But, likewise, it's reasonable of Jones to carry an umbrella if he has in mind to return it to its owner. Since either motivation would rationalize the way that Jones behaved, his having behaved that way is compatible with either interpretation. This is, in fact, overwhelmingly the general case: There are, almost always, all sorts of interests that would rationalize the kinds of behavior that a creature is observed to produce. What's needed to make it decisive that the creature is interested in Y is that it should produce a kind of behavior that would be reasonable *only* given an interest in Y. But such cases are vanishingly rare, since if an interest in Y would rationalize doing X, so too would an interest in doing X. A concern to propagate one's genes would rationalize one's acting to promote one's children's welfare; but so too would an interest in one's children's welfare. Not all of one's motives *could* be instrumental, after all; there must be some things that one cares for just for their own sakes. Why, indeed, mightn't there be quite a few such things? Why mightn't one's children be among them?

The literature of psychological Darwinism is full of what appear to be fallacies of rationalization: arguments where the evidence offered that an interest in *Y* is the motive for a creature's behavior is primarily that an interest in *Y* would rationalize the behavior if it were the creature's motive. Pinker's book provides so many examples that one hardly knows where to start. Here's Pinker on friendship: "Once you have made yourself valuable to someone, the person becomes valuable to you. You value him or her because if you were ever in trouble, they would have a stake—albeit a selfish stake—in getting you out. But now that you value the person, they should value you even more ... because of your stake in rescuing him or her from hard times.... This runaway process is what we call friendship" (508–509).

And here's Pinker on why we like fiction: "Fictional narratives supply us with a mental catalogue of the fatal conundrums we might face some-day and the outcomes of strategies we could deploy in them. What are the options if I were to suspect that my uncle killed my father, took his position, and married my mother?" (543) Good question. Or what if it turns out that, having just used the ring that I got by kidnapping a dwarf to pay off the giants who built me my new castle, I should discover that it is the very ring that I need in order to continue to be immortal and rule the world? It's important to think out the options betimes, because a thing like that could happen to anyone and you can never have too much insurance.

At one point Pinker quotes H. L. Mencken's wisecrack that "the most common of all follies is to believe passionately in the palpably not true." Quite so. I suppose it could turn out that one's interest in having friends, or in reading fiction, or in Wagner's operas, is really at heart prudential. But the claim affronts a robust, and I should think salubrious, intuition that there are lots and lots of things that we care about simply for them-selves. Reductionism about this plurality of goals, when not Philistine or cheaply cynical, often sounds simply funny. Thus the joke about the lawyer who is offered sex by a beautiful woman. "Well, I guess so," he replies, "but what's in it for me?" Does wanting a beautiful woman—or, for that matter, just a good read—really require a further motive to explain it? Pinker duly supplies the explanation that you wouldn't have thought you needed: "Both sexes want a spouse who has developed nor-mally and is free of infection.... We haven't evolved stethoscopes or tongue depressors, but an eye for beauty does some of the same things.... Luxuriant hair is always pleasing, possibly because ... long hair implies a long history of good health" (483–484).

Much to his credit, Pinker does seem a bit embarrassed about some of these consequences of his adaptationism, and he does try to duck them. "Many people think that the theory of the selfish gene says that 'animals try to spread their genes.' This misstates ... the theory. Animals, including most people, know nothing about genetics and care even less. People love their children not because they want to spread their genes (consciously or unconsciously) but because they can't help it.... What is selfish is not the real motives of the person but the metaphorical motives of the genes that built the person. Genes 'try' to spread themselves [sic] by wiring animals' brains so the animals love their kin ... and then the[y] get out of the way" (401). This version sounds a lot more plausible; strictly speaking, nobody has as a motive ("conscious or unconscious") the proliferation of genes after all. Not animals, and not genes either. The only motives there really are, are the ones that everybody knows about; of which love of novels, or women, or kin are presumably a few among a multitude.

Right on. But, pace Pinker, this reasonable view is not available to a psychological Darwinist. For to say that the genes 'wire animals' brains so the animals love their kin" and to stop there is to say only that loving their kin is innate in these animals. That reduces psychological Darwinism to mere nativism; which, as I remarked above, is common ground to all of us rationalists. The difference between Darwinism and mere nativism is the claim that a creature's innate psychological traits are adaptations; that is, that their role in the propagation of the genes is what they're for and is what makes their structure intelligible. Take the adaptationism away from a psychological Darwinist, and he has nobody left to argue with except empiricists.

It is, then, adaptationism that makes Pinker's and Plotkin's kind of rationalism special. Does this argument between rationalists really matter? Nativism itself clearly does; everybody cares about human nature. But I have fussed a lot about the difference between nativism and Darwinism, and you might reasonably want to know why anyone should care about that.

For one thing, nativism says there is a human nature, but it's the adaptationism that implies the account of human nature that sociobiologists endorse. If, like me, you find that account grotesquely implausible, it's perhaps the adaptationism rather than the nativism that you ought to consider throwing overboard. Pinker remarks that "people who study the mind would rather not have to think about how it evolved because it would make a hash of cherished theories.... When advised that [their] claims are evolutionarily implausible, they attack the theory of evolution rather than rethinking the claim" (165). I think this is exactly right, though the formulation is a bit tendentious. We know—anyhow we think we know—a lot about ourselves that doesn't seem to square with the theory that our minds are adaptations for spreading our genes. The question may well come down to which theory we should give up. Well, as far as I can tell, if you take away the bad argument that turns on complexity, and the bad argument from reverse engineering, and the bad arguments that depend on committing the rationalization fallacy, and the atrociously bad arguments that depend on preempting what's to count as the "scientific" (and/or the biological) world view, the direct evidence for psychological Darwinism is very slim indeed. In particular, it's arguably much worse than the circumstantial evidence for our intuitive, pluralistic theory of human nature. It is, after all, our intuitive pluralism that we use to get along with one another. And I have the impression that, by and large, it works pretty well.

So things could be worse, sociobiologists to the contrary not withstanding; I do hope you find that consoling. Have a nice millenium.

References

Allen, C. and Bekoff, M. 1995. "Biological Function, Adaptation, and Natural Design," *Philosophy of Science* 62 (4): 609–622.

Berlin, B. and Kay, P. 1969. *Basic Color Terms: Their Universality and Evolution.* Berkeley: University of California Press.

Block, N. 1978. "Troubles with Functionalism," in C. W. Savage (ed.), *Minnesota Studies in the Philosophy of Science,* Vol IX. Minneapolis, MN: University of Minnesota Press.

Block, N. 1993. "Holism, Hyper-Analyticity, and Hyper-Compositionality," *Mind and Language* 8: 1–26.

Block, N. and Fodor, J. 1972. "What Psychological States Are Not," *Philosophical Review* 81: 159–181.

Carruthers, P. 1996. *Language, Thought, and Consciousness.* Cambridge: Cambridge University Press.

Churchland, Patricia. 1987. "Epistemology in the Age of Neuoscience," *Journal of Philosophy* 84 (10): 544–555.

Churchland, Paul. 1995. *The Engine of Reason, The Seat of the Soul.* Cambridge, MA: MIT Press.

Davidson, D. 1980. "Mental Events," in D. Davidson, *Essays on Actions and Events.* Oxford: Clarendon Press.

Dawkins, R. 1976. *The Selfish Gene.* Oxford: Oxford University Press.

Dawkins, R. 1996. *Climbing Mount Improbable.* New York: Viking.

Dennett, D. C. 1978a. "Intentional Systems," in D. C. Dennett, *Brainstorms,* q.v.

Dennett, D. C. 1978b. *Brainstorms.* Cambridge, MA: Bradford Books/MIT Press.

Dennett, D. C. 1990. "Teaching an Old Dog New Tricks," *Behavioral and Brain Sciences* 13: 76–77.

Dennett, D. C. 1991. "Real Patterns," *Journal of Philosophy* 87: 27–51.

Dennett, D. C. 1995. *Darwin's Dangerous Idea.* New York: Simon and Schuster.

Devitt, M. 1996. *Coming to Our Senses.* Cambridge: Cambridge University Press.

Eldredge, N. 1995. *Reinventing Darwin.* New York: John Wiley.

Elman, J. et al. 1996. *Rethinking Innateness: A Connectionist Perspective on Development.* Cambridge, MA: MIT Press.

Fodor, J. 1974. "Special Sciences," *Synthese* 28: 97–115.

Fodor, J. 1990. "A Theory of Content, Part I," in J. Fodor, *A Theory of Content and Other Essays.* Cambridge, MA: MIT Press.

Fodor, J. 1991. "You Can Fool All of the People Some of the Time, Everything Else Being Equal: Hedged Laws in Psychological Explanations," *Mind* 100 (1): 19–34.

Fodor, J. and Lepore, E. 1992. *Holism: A Shopper's Guide.* Oxford: Blackwell.

Fodor, J. and Pylyshyn, Z. 1988. "Connectionism and Cognitive Architecture: A Critical Analysis," *Cognition* 28: 3–71.

Gallistel, C. 1970. *The Organization of Learning*. Cambridge, MA: MIT Press.

Gleitman, H. 1995. *Psychology*, fourth edition. New York: W. W. Norton & Co.

Gould, S. J. and Lewontin, R. C. 1979. "The Spandrels of San Marco and the Panglossian Paradigm: A Critique of the Adaptationist Program," *Proceedings of the Royal Society of London* B 205: 581–598.

Hempel, C. G. 1965. "The Logic of Functional Analysis," in C. G. Hempel, *Aspects of Scientific Explanation*. New York: Macmillan.

Kamp, H. and Partee, B. 1995. "Prototype Theory and Compositionality," *Cognition* 57: 129–181.

Karmiloff-Smith, A. 1992. *Beyond Modularity: A Developmental Perspective on Cognitive Science*. Cambridge, MA: MIT Press.

Kim, J. 1992. "Multiple Realization and the Metaphysics of Reduction," *Philosophy and Phenomenological Research* 52: 1–26.

Kim, J. 1993. *Supervenience and Mind*. Cambridge: Cambridge University Press.

Loewer, B. and Rey, G. (eds). 1991. *Meaning in Mind: Fodor and His Critics*. Oxford: Blackwell.

Macdonald, C. and Macdonald, G. (eds.). 1995. *Connectionism*. Cambridge, MA: Blackwell.

Marcus, G. F., Pinker, S., Ullman, M., Hollander, M., Rosen, T. J., and Xu, F. 1992. "Overregularization in Language Acquisition," *Monographs of the Society for Research in Child Development* 57: 81–85.

McCauley, R. (ed.). 1995. *The Churchlands and Their Critics*. Cambridge, MA: Blackwell.

McDowell, J. 1994. *Mind and World*. Cambridge, MA: Harvard University Press.

Millikan, R. 1984. *Language, Thought, and Other Biological Categories*. Cambridge, MA: MIT Press.

Mithen, S. 1996. *The Prehistory of Mind*. Thames and Hudson.

Moore, A. W. 1997. *Points of View*. Oxford: Oxford University Press.

Orr, H. A. 1996. "Dennett's Dangerous Idea," *Evolution* 50 (1): 467–472.

Peacocke, C. 1992. *A Study of Concepts*. Cambridge, MA: MIT Press.

Pinker, S. 1997. *How the Mind Works*. New York: Norton.

Plantinga, A. (forthcoming). "Naturalism Defeated." Unpublished manuscript, University of Notre Dame.

Plotkin, H. 1997. *Evolution in Mind*. London: Alan Lane.

Putnam, H. 1967. "The Nature of Mental States," in D. Rosenthal (ed.), q.v.

Pylyshyn, Z. 1984. *Computation and Cognition: Toward a Foundation for Cognitive Science*. Cambridge, MA: MIT Press.

Rosch, E. H. 1973. "Natural Categories," *Cognitive Psychology* 4: 328–350.

Rosenthal, D. (ed.). 1991. *The Nature of Mind*. Oxford: Oxford University Press.

Savin, H. and Bever, T. G. 1970. "The Nonperceptual Reality of the Phoneme," *Journal of Verbal Learning and Verbal Behavior* 9: 295–302.

Schiffer, S. 1991. "Ceteris Paribus Laws," *Mind* 100 (1): 1–17.

Smolensky, P. 1987. "The Constituent Structure of Mental States: A Reply to Fodor and Pylyshyn," *Southern Journal of Philosophy* 26: 137–160.

Smolensky, P. 1988a. "On the Proper Treatment of Connectionism," *Behavioral and Brain Sciences* 11: 1–23.

Smolensky, P. 1990. "Tensor Product Variable Binding and the Representation of Structure in Connectionist Systems," *Artificial Intelligence* 46, 159–216.

Smolenksy, P. 1995a. "Connectionism, Constituency, and the Language of Thought," in Macdonald and Macdonald, q.v.

Smolensky, P. 1995b. "Reply: Constituent Structure and Explanation in an Integrated Connectionist/Symbolic Cognitive Architecture," in Macdonald and Macdonald, q.v.

Sober, E. 1984. *The Nature of Selection*. Cambridge, MA: MIT Press.

Index